Sport Science

Also of interest:
The Physics of Ball Games   C. B. Daish

Sport Science

Roy Hawkey
Head of Science, Bohunt School, Liphook, Hants.

HODDER AND STOUGHTON
LONDON SYDNEY AUCKLAND TORONTO

British Library Cataloguing in Publication Data

Hawkey, Roy
 Sport science.
 1. Physical education and training – Great Britain
 I. Title
 613.7'07'1241   GV245

ISBN 0 340 25127 1

First printed 1981

Copyright © 1981 R.S. Hawkey

All rights reserved. No part of this publication may be reproduced or transmitted in any form or by any means, electronic, or mechanical, including photocopy, recording, or any information storage and retrieval system, without permission in writing from the publisher.

Printed in Great Britain for Hodder and Stoughton Educational, a division of Hodder and Stoughton Ltd., Mill Road, Dunton Green, Sevenoaks, Kent, by William Clowes (Beccles) Limited, Beccles and London

Preface

This book arose from a CSE mode 3 course in Sport Science. It can be used in this way, or as part of a course in Physical Education or Science. It may also serve as an introduction to the subject at higher levels.

It is unlikely that the arrangement of material in the book will suit everybody who uses it. Textual cross-references and a comprehensive index have been provided to facilitate any particular approach. To have ordered the material either by sport or by scientific principle was felt to require too much repetition or lack of continuity. The result is a compromise.

Although mathematics is an essential scientific tool, calculations are needed at this level mainly for their results. Most calculations are therefore given in a form suitable for direct entry into an electronic calculator. Such calculations are printed vertically and contain factors necessary to convert units. The following notation is used:

notation	meaning	alternative
M	Memory	STO
C	clear display	CLR
MR	recall memory	RCL

Questions are provided at the end of each chapter. These largely recap the material of the chapter, although some do require further thought. Appendix C contains longer questions, covering greater breadth and depth of material. Many of these have been used in Mode 3 CSE papers administered by the South Western Examinations Board.

Most chapters contain an 'Extra' section. This develops further ideas introduced in the chapter, or contains more difficult material.

Contents

Preface v

1	What is Sport Science?	1
1 Extra	Odds and Ends	9
2	How do I Measure up?	10
2 Extra	Race and Sex	26
3	What to do with Data	28
3 Extra	A Closer Look at Records	37
4	Analysing Movement	39
4 Extra	Left and Right Hand. Momentum	49
5	A First Look at Ball Games	51
6	More about Ball Games	58
6 Extra	A Sport Science Advertisement	64
7	How the Body Reacts	65
8	What is Fitness?	76
8 Extra	Efficiency	88
9	Athletics	90
9 Extra	Technology in Athletics	97
10	Mechanical Sports	98
10 Extra	Man-powered Flight	104
11	Playing Surfaces	106
11 Extra	Playing Surfaces and the Weather	111
12	Practice, Training, Injuries	112
12 Extra	Smoking and Fitness	120
13	Sport and Social Science	122
13 Extra	Floodlights	127
References and Further Reading		128
Appendix A	A Syllabus for CSE Examinations	129
Appendix B	Ideas for Project Work	131
Appendix C	Examination questions	132
Index		136
Acknowledgements		138

1 What is Sport Science?

Sport Science is the science of sport!

It is the application of scientific ideas to sporting activities. It is an essential part of physical education, and can help answer the questions which may lie behind the desire to improve performance.

> Am I the right build for my sport?
> How can I use information about sport?
> How can I improve my techniques?
> Why do some balls bounce higher than others?
> What is top-spin?
> Can I react quickly enough?
> How powerful are my muscles?
> What are the key factors in athletics?
> Where do the forces act on a cyclist?
> How much effect do playing surfaces have?
> How effective is practice?
> Has social science anything to offer?

These are the kinds of question which this book will try to answer. Or, at least, it will show the problems in more detail, and give some ideas about how they might be solved. Each of the questions above is dealt with, at least to some extent, in the chapters which follow.

In this first chapter we shall look at the basic question, 'How do we find out?'

A Sport Science Experiment

The sport scientist, like other kinds of scientist, must do experiments if he is to make discoveries. The ideas for these experiments may come from many different places: first-hand observation, articles in newspapers, remarks by commentators, or odd questions.

Why, for example, have many male swimmers taken to wearing swimming-hats? Or even shaving their heads completely?

Is it just that long hair gets in the way? Or does the very presence of hair affect the body's flow through the water? If it does affect the flow, we ought to be able to measure a difference between swimming with a hat, and swimming without one.

But how do we set up such an experiment? Here's a discussion among a group of students about this problem and all the factors and difficulties involved:

ALAN: Can we just find two swimmers, one who always wears a hat, and one who never does? Then we could compare their performances.

BRENDA: No—there'll be too many other differences. We might choose a complete beginner and an Olympic champion! We can only look at one difference: hat or no hat.

CAROL: Perhaps we should use two *groups* of people, and average

	their results. Is this any good?
MIKE:	No, again. It is better, though; individual differences will be smaller. But it still could be that all the good swimmers were in one group.
TONY:	Let's *make* our swimmers wear hats—or not—rather than let them choose for themselves.
BRENDA:	That's better. Now we can rule out the possibility that choice may be something to do with ability. Perhaps coaches tell their best pupils to wear hats, or something.
SUE:	Well, just in case, what if we choose our two groups at random, ...
LEE:	.... and make one group swim with hats first, then the other?
MIKE:	Good idea. Can we go any further, make any more improvements?
JILL:	Could we compare each group's performance, with and without hats?
SAM:	Why stop at 2 groups, why not 3? Or 10?
MIKE:	We've only got a couple of hours, Sam. And, anyway, who's going to work out the results? You?
BRENDA:	Isn't there another way, with just one swimmer? Get him (or her) to swim alternately with and without hat, then compare the times. What's wrong with that?
TONY:	That seems too easy. What do you think, Graham?
GRAHAM:	It might seem to be the best way of avoiding the individual differences. But the very fact that people are different means that any results might not be true for anyone else. So, the group method is much better. Scientists would call the two groups

> 'experimental'—with the hats, and
> 'control'—without, 'normal'.

Changing them over, like Lee suggests, is a good way of ruling out any other factors.

It's always important in science to be sure that we are measuring what we think we are measuring. Then we can make judgements and recommendations. If you think about them carefully, you will find that some of the experiments in this book are not perfect!

But, of course, it's not always possible to do experiments to answer questions. Often, we have to rely simply on looking, on observation. We can do this by collecting facts.

Collecting Information

Much of the basic, unglamorous work of science is concerned with collecting information.

Information can be collected from many different places. One technique

is to use a tally-chart while watching a sporting event. In this example, a list was made of events which happened during a rugby union match, and a check made each time one took place. Part of the original record looked like this:

| Free kicks | ⟊⟊⟊ ΙΙ   7 | ΙΙΙΙ   4 |

The final, completed figures were:

	IRELAND	ENGLAND
Scrums won	12	10
Scrums 'against the head'	3	0
Line-outs win	22	28
Free kicks	7	4
Penalties	8	10
Attempts at goal	1	4
Penalty goals	1	1
Tries	1	1
Conversions	1	0
Dropped goals	1	0

'Fact-finder' from a rugby union international

This sort of 'fact-finder' is often used by commentators. It can be used to record various events in a variety of sports, for example:

    tennis : double faults, aces, first serves 'in'
    soccer : corners, free-kicks
    golf   : holes 'birdied'
    cricket : boundaries hit

Having got the information we should, as scientists, try to explain it, with a hypothesis or theory. Then we test our ideas by trying to predict future events, either in an experiment, or in the real world. For example, we might notice that sport was disrupted by very cold winters in 1947, 1963 and 1979. Rather like an IQ test, we predict that the same 16-year interval will make 1995 also a bad winter:

    1947   1963   1979   1995?   ?   ?

Sometimes, to check a theory, we don't have to wait so long. We can use information that we already have. Football supporters often say that matches between nearby clubs (local 'derbies') are often draws. They suggest various reasons for this, such as reduced travelling, more supporters and so on. It is quite easy to investigate the 'prediction' that more of these 'derby' games are drawn:

	Number	Draws	%
Local 'derbies'	48	14	29
Others	414	118	29

Data from League Division I, 1977-78.

Are local 'derbies' more often drawn?

This might, the scientist would say, be a freak result. So we had better check other results, from other seasons and other divisions. Some other examples are given.

Fixture	Division	H	A	D
Norwich v Ipswich	1	5	3	3
Bradford C. v Barnsley	4	5	1	5
Rangers v Celtic	Sc. Prem.	9	3	5
Darlington v Doncaster	4	5	4	3
Liverpool v Everton	1	4	3	6

Other 'derby' results 1972-1978/79

Predicting in Science and Sport

Scientists try to predict the result of experiments by finding a pattern in their knowledge. Sports enthusiasts try to do the same with results. This is not in itself scientific, but some of the techniques used are of use to the scientist. And, of course, millions of pounds are spent in trying to predict results of horse races and soccer matches.

To know that in 1978-79 there were

        827 home wins,
        429 away wins,
and 521 draws

may not appear to give much help. But, over a period of time, the ratio of home:away:draw is about 2:1:1. So, forecasts tend to be in this sort of ratio. Use of ratios is important in many aspects of science.

Team line-ups are normally symmetrical:

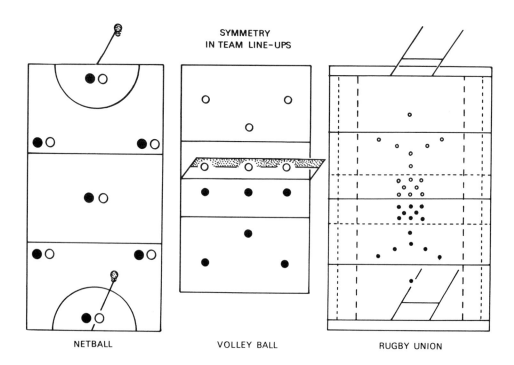

SYMMETRY IN TEAM LINE-UPS

NETBALL   VOLLEY BALL   RUGBY UNION

Results, too, may have symmetry. Look at this record:

Played	Points	Home			Away			Goals	
		W	D	L	W	D	L	F	A
41	40	10	6	5	4	6	10	56	58

No symmetry here. But what if the team won its final match away by 2 goals to nil?

Questionnaires

Some sorts of question cannot be answered either by experiment or observation. We must ask. Not just odd questions, though. If we want answers which mean anything, we need properly designed questionnaires. We need the sort of thing used in opinion polls and market research. Some examples are given here.

Check-lists are a simple type:

'Tick which of these activities you think are sports:'

hockey	☐	gymnastics	☐	tennis	☐
chess	☐	soccer	☐	fishing	☐
boxing	☐	skating	☐	golf	☐

There are also questions which include an element of choice:

'Tick to show how often you play these sports:'

	Frequently	Sometimes	Never
Soccer			
Netball			
Hockey			
Squash			

Or this kind:

'Which is nearest to your view of sports centres?'

    a waste of money
    keep people off the streets
    make no difference
    a good idea
    every town should have one

If the questionnaire asks for basic personal data, such as age, sex, level of fitness, you can compare the answers from different groups of people.

Sorting out the results of a survey can be as hard as asking the right questions. For example, what does this tell you?

	Boys	Girls
Play tennis	22	27
Do not	28	23

Can we say that more children do not play tennis (51) than do (49)? Or were we just lucky with our sample? If the next three people asked were players, this might suggest the opposite. Similarly, can we state that more girls really do play tennis? Could this be a chance result?

The statistics needed to answer this are more than can be dealt with here. What the statistical test would tell is that the result is *not* at all significant. In this case there is no real boy/girl difference.

Sport, Pastime or Art?

Here is a list of recreational activities:

archery	crosswords	hockey	soccer
ballet	dancing	painting	swimming
bird-watching	fishing	sailing	tennis
chess	golf	show-jumping	volleyball
cricket	hill-walking	skittles	weight-lifting

Part of science is concerned with classifying things. As sport scientists we should be able to classify these activities into sports and pastimes. To do so we need to know what exactly we mean by the term 'sport'.

Make three lists from the activities above:

(a) those which definitely are *not* sports;
(b) those which definitely *are* sports;
(c) those which *may* or *may not* be sports.

In your first list you will have bird-watching and painting. In list (b) will be golf and tennis. But what about fishing? Is that a sport or not? It depends on what we really mean by 'sport'; it depends on exactly how we define the word.

You can, of course, simply look it up in the dictionary. But a scientist is more likely to produce a set of rules. He uses these to judge each activity. Only if an activity obeys all the rules can it be classified as a sport.

Now the problem is to write the rules. One example is given: you can argue about these, and try to produce your own set of rules.

A sport

(a) involves physical activity;
(b) is competitive;
(c) contains a directly measured quantity, e.g. time, goals, points, distance.

Try these rules on the activities in the list.

Would your three lists have been different?

What happens to gymnastics, ice-skating, fishing and jogging if these rules are applied?

Could you change any of the rules without having to say that a disco dancing contest was a sport?

Classifying Sports

Can you think of a way of putting sports into groups? Of classifying them? This should help us understand more, as the same scientific principles may well apply to similar kinds of activity. For example, tennis, squash and badminton have obvious similarities.

One way of classifying sports is:

(a) athletic sports
(b) ball games
(c) 'combat' sports
(d) 'target' sports
(e) water sports
(f) winter sports

Systems like this, though, tend to either miss out some sports, or put some in twice. This scheme, for example, can't include badminton. But, if we add 'racket games' as another group, tennis and squash will occur twice, as they are also ball games. The only way to fully classify sports is to use the method of sets, with a Venn diagram.

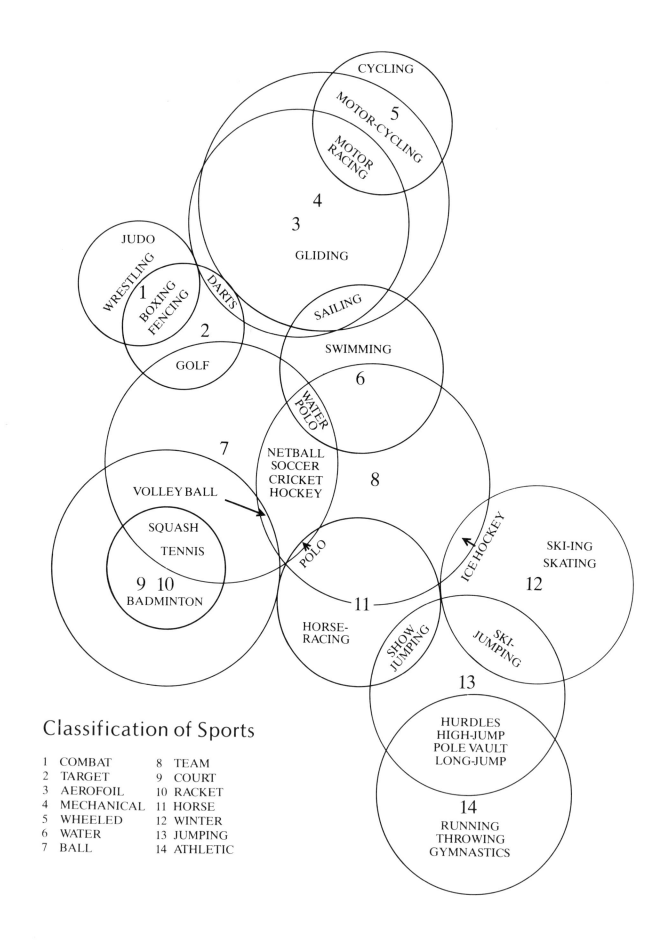

Questions

1. In an experiment, runners wearing a new kind of track shoe are compared with runners wearing ordinary spikes. Which is the 'control' group, and which the 'experimental'?
2. Make up a suitable tally-chart for hockey, basketball or netball.
3. Write 3 questions suitable for a questionnaire on sport.
4. Suggest 3 rules for deciding whether an activity is a sport.
5. Which sport is (a) target, aerodynamic?
   (b) ball, team, court?
   (c) team, water, ball?
   (d) winter, team?
   (e) court, racket, but not ball?
6. Which sport is (a) a target ball game?
   (b) an 'aerofoil' water sport?
   (c) a winter sport played in teams?

Extra: Odds and Ends

Odd Man Out

Have you ever done one of those IQ tests, where some of the questions ask, 'Which is the odd one out of . . .?'? Twenty years ago, when these tests were very popular in schools, there was lengthy correspondence in the 'Times' on a sports 'odd one out' question. The four sports to choose from were:

        billiards   cricket   hockey   soccer

There was lengthy argument about which was the 'best' answer. What do you think? Can you make out a good case for each of them?

The series of letters include answers like:

- billiards : more than one ball, played on a table;
- cricket : ball not put into any kind of net;
- soccer : no 'hitter' used to propel the ball.

One correspondent argued that hockey must be the right answer as he couldn't think of a reason for it to be the odd one! Yet, of course, hockey is the only one listed which does not have professional players—except for ice-hockey.

Keys

Scientists, particularly biologists, often use keys to help them identify and describe things. The 'selectagraph' (page 31) is a kind of key. The discussion of odd one out can help us to make a key to these four sports.

1	Uses a 'hitter'	2
	No hitter used	SOCCER
2	One ball only	3
	More than one ball	BILLIARDS
3	Ball hit into net	HOCKEY
	Ball not hit into net	CRICKET

Can you see how the key works?

Start at '1'. Choose the line which fits the sport you are thinking of. At the end of the line is either an answer, or another number. If there is another number, move to this number, and repeat the process. When you reach a name, you have the answer.

Below is a similar key for football teams. It needs to be revised every season, as teams change. This is not as easy as it seems, for you can't just replace, say Luton Town with Wimbledon, unless their kit happens to be identical.

Key to 1st Division Football Teams 1979-80

1.	Shirts and shorts same colour	2
	Shirts and shorts different	6
2.	White	LEEDS UNITED
	Not white	3
3.	Blue	4
	Red	5
4.	With stripes	COVENTRY CITY
	Without	MANCHESTER CITY
5.	Sleeves plain	LIVERPOOL
	Sleeves with stripes	MIDDLESBROUGH
6.	Shirts striped	7
	Not striped	11
7.	3 colours	CRYSTAL PALACE
	2 colours	8
8.	Red and white	9
	Blue and white	10
9.	Black shorts	SOUTHAMPTON
	White shorts	STOKE CITY
10.	Wide stripes	WEST BROMWICH ALBION
	Narrow stripes	BRIGHTON & HOVE ALBION
11.	Sleeves different from rest of shirt	12
	Sleeves same as rest of shirt	13
12.	Sleeves white	ARSENAL
	Sleeves blue	ASTON VILLA
13.	Shirts red	14
	Not red	16
14.	Socks black	MANCHESTER UNITED
	Socks red	15
15.	Badge a tree	NOTTINGHAM FOREST
	Badge not a tree	BRISTOL CITY
16.	Shirts blue	17
	Not blue	18
17.	Badge with animal	IPSWICH TOWN
	Badge not so	EVERTON
18.	Shirts white	19
	Shirts yellow	21
19.	Badge ram or cockerel	20
	Badge not a ram or cockerel	BOLTON WANDERERS
20.	Badge a ram	DERBY COUNTY
	Badge a cockerel	TOTTENHAM HOTSPUR
21.	Shorts black	WOLVERHAMPTON WANDERERS
	Shorts green	NORWICH CITY

2 How do I Measure up?

Most people have some idea of what their 'ideal' sportsman would be like, in terms of build, strength, temperament and so on. Such views will be affected by thoughts of particular sports. Jockeys and basketball players need rather different body designs.

In this chapter, we will look at some of the physical and psychological factors which make up a sportsman. We will see how they can be measured, and, with our own set of results, decide at which sports each should be best—at least, in theory.

Muhammed Ali

Mass and Weight

To scientists, 'mass' and 'weight' have different meanings. Mass is how much of an object (or person) there is; weight is the force with which this mass pushes down on to the Earth. Weight should be recorded in newtons because it is a measure of force; mass is measured in kilograms.

It will be useful to know your own body mass. You can find it by standing on platform scales, reading in kilograms (kg).

Make a note of your body mass. You will need to refer to it again, and you will also be able to compare yourself with top-class sportsmen and women, as well as with your friends and others of your age. (See Fig. 2.1 on page 15.)

Lester Piggott

Body Mass in Sport

Body mass is an important factor in any sport. For some sports it is so

Vassili Alexeev

Alberto Juantorena

Tamara Press

David Starbrook

Laurie Cunningham

vital that there are rules about it. For others, there may be a definite advantage in being heavy. There are even some sports, or certain parts of them, where lightness is an advantage.

From the sports given, make lists of those which

(a) have rules about body mass,
(b) are best for heavy people,
(c) are best for light people,
(d) are not affected by body mass.

Archery	Ice hockey
Athletics : distance running	Judo
high jump	Netball
long jump	Rowing : oarsman
sprinting	coxswain
throwing	Rugby : forward
Badminton	back
Basketball	Skating
Boxing	Skiing
Canoeing	Soccer
Cricket	Squash
Darts	Swimming : long distance
Fencing	sprinting
Golf	Table tennis
Gymnastics	Tennis
Hockey	Volley ball
Horse riding : racing	Weightlifting
show-jumping	Wrestling

You will probably have found few sports in which body mass by itself is particularly important, if we ignore the factor of height or general build. Only the 'combat' sports and weightlifting actually have rules about it, to prevent unfair competition and injury.

The sportsmen and woman shown have body masses of 70, 84, 100 and 118 kg. Can you tell which is which?

Height

Height, like mass, can be an important factor in sport. It is fairly easy to list those sports where being very tall is an advantage, and others where it is an advantage in certain positions. But, however many short sportsmen there may be, it is difficult to think of sports where being short—as opposed to light—is a real advantage. And there are no sports which have rules about height.

Choose from the sports given before and make a list of those for which it is best to be

(a) tall,
(b) tall, for certain positions only,
(c) short.

Steve Assinder

Tony Greig

Valeri Borzov

Jimmy Connors

Willie John McBride

As with mass, it will be useful to know your own height. You will need someone to help you, to make a mark on the wall or blackboard, and a metre rule or two.

The sportsmen shown have heights of 1.78, 1.80, 1.96, 2.02 and 2.16 metres (m). Try to match each with his correct height.

Height and Mass

The graphs (Fig. 2.1) will give you some idea of how you measure up in terms of height and mass, compared with others of your age. Make sure you look at the right graph—male or female.

Are you above, below or average mass for your height?

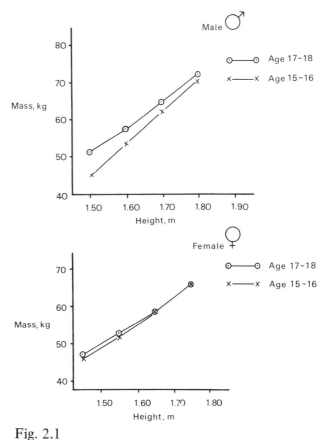

Fig. 2.1

Remember that these are mean (average) values: you are not peculiar if you are not exactly average.

Centre of Gravity

Centre of gravity can be thought of as the point at which gravity acts on a body—human or object. It's where the weight is reckoned to be concentrated. For a regular, symmetrical object, such as a ball, the

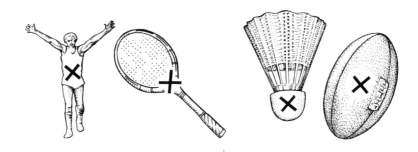

centre of gravity is its actual centre. For irregular things like human beings, shuttlecocks and tennis rackets its position may not be so obvious.

To find the centre of gravity of a simple object, such as a piece of card, is fairly easy:

(1) hang it from a single pivot—a pin through it,
(2) hang a 'plumbline' from the same pivot,
(3) mark this vertical line on the card,
(4) repeat 1–3 using a different position.

Both lines will pass through the centre of gravity; it is where they cross.

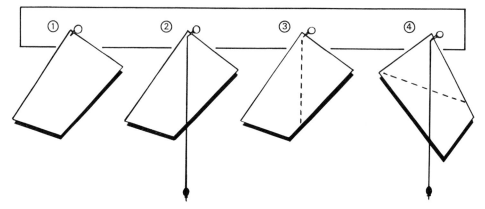

This method will work for solid objects, as long as they are not totally irregular. Be warned, though, that the centre of gravity of an object can lie in space, and not in the object at all!

To find the centre of gravity of a person we don't hang him from a pin, but lie him on a plank. Four measurements are needed to find the height of the centre of gravity from the feet; we assume that it is otherwise exactly in the centre.

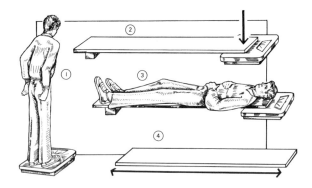

Measure:

(1) the weight of the person alone—this is found by standing on platform scales marked in newtons (N);
(2) the force of the plank, supported on a brick and the newton-meter scales (K);
(3) the force exerted by the person lying on the plank, feet over the brick (P);
(4) the length of the plank, in metres (L).

The distance of the centre of gravity above a person's feet is:

$$\frac{L(P-K)}{N} \qquad P - K = \times L \div N =$$

The formula on the right shows how this can be entered directly into a calculator. This gives the distance (in metres) above the feet of the body's centre of gravity. It is normally a little above the navel.

The position of the centre of gravity is important in two ways. One is balance.

Anything, a person included, will balance if its centre of gravity is directly above its base; the force of gravity will act directly down so that the weight is fully on the base:

As soon as the centre of gravity moves away from its stable position above the base, the force of gravity will act more on one side than the other. The object will turn, and fall: it has become unstable.

The lower the centre of gravity, the more likely it is to stay above the base; and the larger the base, the more stable the object. This means that short, fat people tend to be more stable (physically, not emotionally) than tall, thin people.

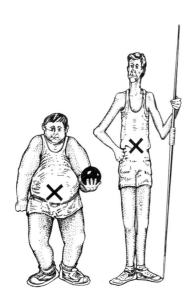

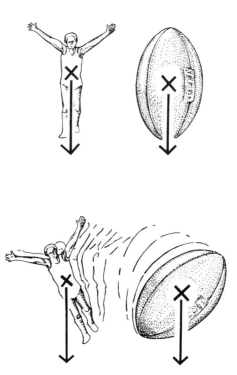

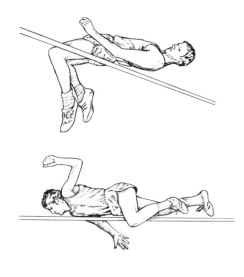

In any sport where keeping upright is important, a low centre of gravity is an advantage.

The second feature of the position of centre of gravity is its effect on work, energy and power. As we shall see later (Chapter 8) the amount of energy required for an activity depends, among other things, on how far the force is applied: twice as much energy is needed to jump 2m as 1m. But the distance we should really measure is the movement of the centre of gravity, for it is here that the force acts.

In comparing high jumps, for example, we should measure the height moved by the centre of gravity of the jumper, not his feet. Look at the drawings of different high jump styles.

You should be able to see which moves the centre of gravity the most, and which the least, in jumping the 'same' height. Dick Fosbury invented the 'Fosbury flop' technique as a way of gaining extra height without doing extra work.

Fosbury himself stands 1.93m tall; most top-class high-jumpers tend to be very tall and thin. This gives them a higher centre of gravity, and therefore less work to do when jumping. One interesting exception to this was the great Russian high jumper Valeri Brumel. The first man to clear the 'barrier' height of 2.13m ('seven feet'), he was only 1.85m tall and a heavy 79kg—heavy, that is, for a high-jumper.

Making Manikin Men

It is possible to measure much more about the body and to calculate more still. Head, trunk and limbs can be measured directly. Centre of gravity can be found for individual body parts. Even the masses of limbs can be estimated, as their proportions are fairly standard.

With enough information we can make a 'manikin' out of card and paper fasteners, which will be a fairly accurate model of the body. The shapes drawn on page 19 will save you a lot of trouble in measuring and calculating, although they are really for the 'standard' 76kg, 1.78m athlete. They are drawn to 1/6 scale. Simply trace them onto card.

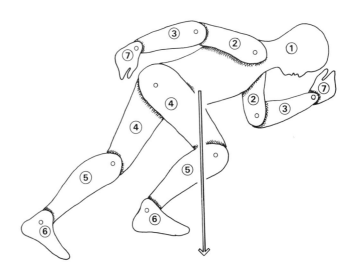

A manikin like this can be used to find out what happens to the body's centre of gravity during an activity. The calculation is rather tedious, but can be done with a calculator.

In the sprint start position shown on page 18, a vertical line through the centre of gravity falls in front of the feet. The runner will therefore tend to fall forwards, but the thrust applied to the ground will keep him balanced. By moving the manikin into any sporting position (by comparison with a photograph) it is possible to find out what effect the position of the centre of gravity is likely to have.

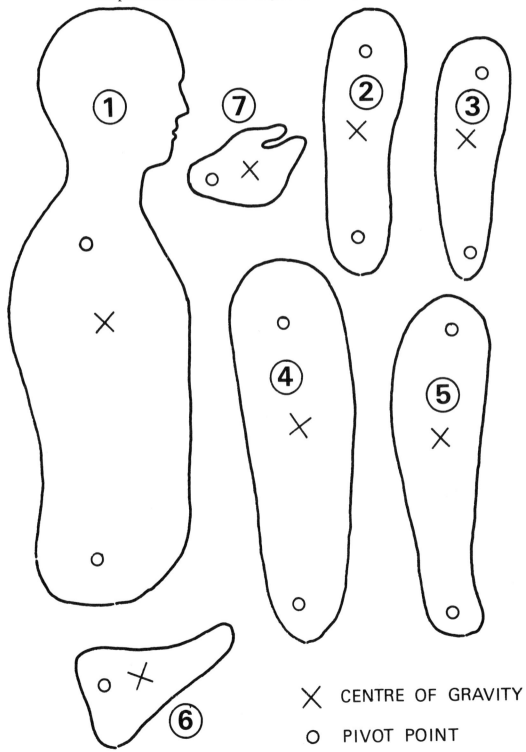

× CENTRE OF GRAVITY

○ PIVOT POINT

Body Temperature

It's obvious that we produce heat whenever we do any sort of physical activity. What is less obvious is what happens to the body temperature at this time. Does it go up? Stay the same? Or does all that sweating make it actually go down?

To try to answer that question we first need to know what the body's temperature is at rest, under 'normal' conditions. To measure it, we use a special 'clinical' thermometer:

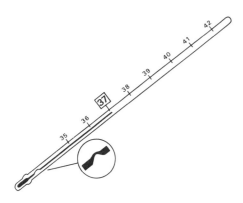

This is kept under the tongue for a minute, then taken out and the temperature read. The kink in the mercury thread means that the thermometer reads the same after it has been taken out of the mouth. (What would happen with an ordinary thermometer?)

37°C is often said to be 'normal' body temperature. What does this mean? Is it really normal?

No. It's the mean (or average) temperature. Yours may be as much as a degree more or less and still be normal—normal, that is, for you. Each person has his or her own normal temperature; the population has an average. It will be useful to record your own body temperature. You can compare it with the mean, and see what happens to it during activity.

Surface Area

One important factor in keeping body temperature steady is the amount of heat lost through the skin. This depends, among other things, on how well insulated the body is (how much fat) and how large the surface area is (how much skin).

You could measure the surface area by wrapping the body in paper, but it is easier to use an equation or a chart.

$$\text{Surface area} = (\text{mass})^{0.425} \times (\text{height})^{0.725} \times 0.00718$$

(This is an empirical—trial and error—equation)

With the equation we can calculate the surface area from the height (in cm. this time) and mass (in kg).

You really need a calculator with a y^x key for this:

mass
y^x
0.425
=
M
C
height
y^x
0.725
=
×
MR
×
0.00718
=

This gives the surface area in sq m (m^2)

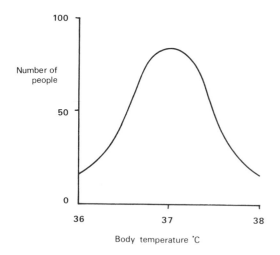

Without such a calculator, use this chart. Line up your mass and height with a ruler, and your surface area is given where it crosses the central scale.

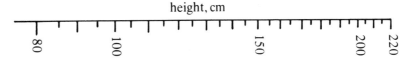

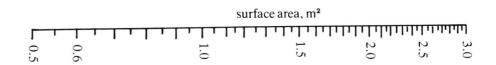

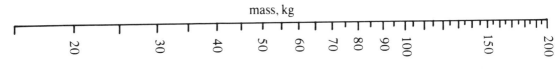

Body Shape

The technical term for body shape is 'somatotype'. There are three basic types:

> endomorph—short and dumpy,
> mesomorph—'medium' build,
> ectomorph—tall and skinny.

Typical somatotypes are shown here:

There are two simple ways to decide which you are:

1. Compare your height and mass with the average for your age (Fig 2.1). If you are taller and lighter, you are probably ectomorphic; shorter and heavier suggests endomorph.

2. Use calipers to measure the amount of fat under the skin. Endomorphs have lots of such subcutaneous fat, ectomorphs very little.

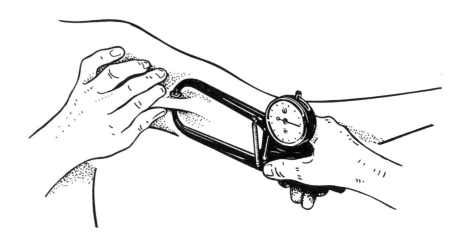

Suitability for Sport

We can now consider the physical factors that affect a person's choice of sport. See if you can choose the best sport (from the symbols given) for each of the sportsmen and women shown.

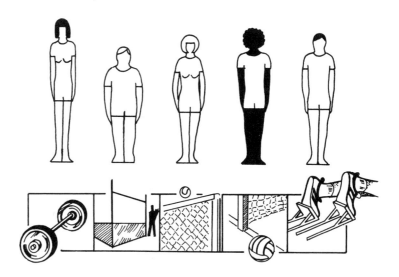

Which sport are you best suited for?

Looking at each feature in turn, rule out any unsuitable sports. The top table on page 23 gives some ideas; see if you can add any more.

Temperament

However easy—or difficult—it may be to find a sport to suit your physique, this is not the only factor to consider. Patience, concentration, dedication, short bursts of aggressive energy: these are some of

Feature	Good for	Poor for
Tall	basketball, cricket, fencing, netball, rugby (forward), volleyball	
Heavy	athletics (throwing), rowing (oarsman), rugby (forward), weightlifting	athletics (distance running), jockey, rowing (coxswain)
High centre of gravity	athletics (jumping), diving	boxing, judo, skiing, weightlifting, wrestling

the things which a potential sportsman must consider. Not only do jockeys and weightlifters need different builds, they also require very different temperaments.

There have been many attempts to produce a 'sporting temperament scale'; an example is given here. Rate each item on a 1–5 scale: 1 if it's near the character on the left, 3 for half-way, 5 for the one on the right. See what sort of pattern you get.

adventurous, imaginative	1 2 3 4 5	conventional
aggressive, forceful	1 2 3 4 5	passive, patient
ambitious	1 2 3 4 5	content
brave	1 2 3 4 5	realistic
careful	1 2 3 4 5	careless
competitive	1 2 3 4 5	submissive
decisive	1 2 3 4 5	hesitant
dedicated	1 2 3 4 5	carefree
enthusiastic	1 2 3 4 5	reluctant
proud	1 2 3 4 5	humble
selfish	1 2 3 4 5	co-operative, open-minded
self-controlled	1 2 3 4 5	undisciplined

Table 2.1

Try it for yourself, and for the 'ideal' sportsman. Can you find features which you would expect to differ from one sport to another?

What Makes Me Like I Am?

The simple answer to this question is 'Bones, muscles and food.'

In this last section of the chapter we shall look at this answer in a bit more detail. Firstly those bones, with all their strange names: without them we'd be nothing but a mass of jelly on the floor. And the muscles too; their names are not much better, but without them our joints would fold up, and we'd be crunchy jelly.

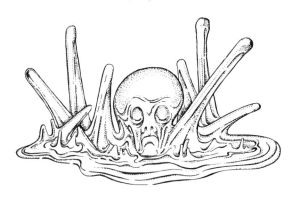

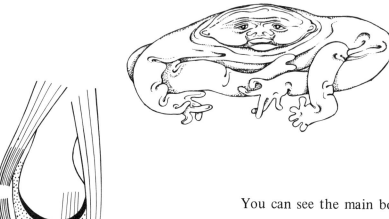

You can see the main bones and muscles—and their names—in Fig. 2.2.

Where the bones and muscles act together—in our joints—there are four other important materials:

    cartilage: prevents friction between bones
    synovial fluid: lubricates ('oils') the joint
    tendons: join muscle to bone (non-elastic)
    ligaments: join bone to bone (elastic)

Knee and hip joints are shown.

On top of the basic structure provided by the bones and muscles, our actual build depends on the balance between food and exercise. Diet is very important in determining physique. Japanese sumo wrestlers eat more than 7kg of steak every day, and just look at the daily menu for Oxford oarsmen in training for the boat race:

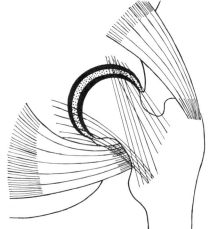

CARTILAGE
FLUID
LIGAMENT
TENDON

Breakfast	Lunch	Dinner
2 bowls cereal 350cm³ orange jce. 4 sausages, bacon, fried bread, tomatoes, sweet tea	Thick soup Pizza Heavy pud.	Pate & toast 350g steak, potatoes, vegetables, Pudding Sweet lemon juice
Extra: up to 10 Mars bars!		

Table 2.2

The poor cox, meanwhile, is lucky if he gets a few pieces of crispbread, and his lemon juice is unsweetened. If this were not enough, he has to

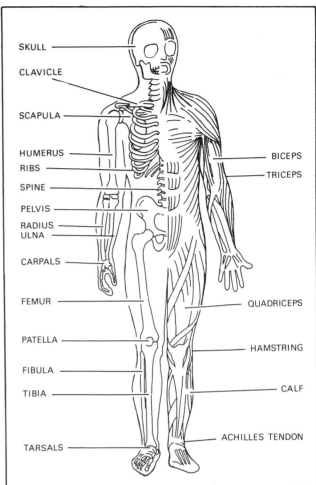

Fig. 2.2

take frequent sauna baths to help lose vital body mass—as much as 10kg in the months before the race. This is cheating though as it's mostly water, which is soon replaced.

Energy taken in as food, but not used, is stored. Normally, it is stored as fat, so, if the body's shape and condition are to be maintained, intake and output of energy must balance. We shall look at this in Chapter 8.

Questions

1. Which unit is used for (a) body mass, (b) body weight?
2. Name three sports in which it is an advantage to be heavy, and two sports for which it is better to be light.
3. Name three sports for which it is best to be tall.
4. Use Fig. 2.1 to find the masses of
   (a) a 15 year old boy, height 1.65m,
   (b) a 17 year old girl, height 1.60m,
   (c) a 16 year old boy, height 1.80m.
5. Why will man A balance, but B and C fall over?
6. Why are high-jumpers usually tall and thin?
7. What is a manikin man? What is it used for?

8 What temperature is shown on the clinical thermometer?
9 Find the surface area of an athlete, height 1.80m (180cm), mass 80kg.
10 Look carefully at the sportsmen shown in this book. List those who are (a) endomorphs, (b) mesomorphs, (c) ectomorphs.
11 (a) Which arm bones are in similar positions to these leg bones? (i) femur, (ii) tibia, (iii) fibula.
    (b) Which leg bone has no 'relative' in the arm?
12 Complete this table to show the difference between ligaments and tendons.

	Ligaments	Tendons
Elastic?		
Joins bones to?		

13 What sort of temperament would suit (a) a golfer, (b) a gymnast, (c) a weightlifter? Use the features in Table 2.1 to help you.

Extra: Race and Sex

Two further factors affect body shape and size. Both race and sex are important, although not always freely discussed.

When Jesse Owens won several gold medals in the 1936 Munich Olympic Games, Hitler was not pleased, for it did not fit his ideas of the 'supreme race'. Now, in the 1980s, it seems quite normal for many of the world's leading sportsmen and women to be black. Indeed, it is thought unusual if the heavyweight boxing champion and Olympic sprint champions are not black. Certainly, in these power events, such racial difference seems to tell, yet top-class black tennis and soccer players are rare. In distance running, however, it is the wispy, high-altitude born Kenyans and Ethiopians who have done exceptionally well in recent years.

Although all races now compete freely against one another, the same is not true of the two sexes. Men are physically stronger than women. Whatever equality may mean in other spheres, it certainly does not apply to most sports. Table 2.3 compares

Event	Male	Female
Running: 100 m	9.95 s	10.88 s
400 m	43.86 s	48.60 s
High Jump	2.43 m	2.01 m
Swimming: 100 m free	49.44 s	55.41 s
200 m breaststroke	2min 15.11 s	2min 28.36 s
Cycling: 500 m	28.6 s	35.0 s

Table 2.3: Male and female records compared

Jesse Owens

Evonne Cawley

Pele

Arthur Ashe

top performances by male and female athletes.

This is not to say that women cannot reach a high standard of skill and performance, but it does explain the need for 'chromosome testing' of women athletes.

Most body cells contain 23 pairs of chromosomes; 22 of these pairs are identical in male and female, but the 23rd pair is either XX or XY. XY produces a male, XX a female. The test involves looking for a Y-chromosome. For this reason, Renée Richards was not allowed to play in certain women's tennis tournaments after a 'sex change' operation, and some shot putters have been disqualified from athletics championships.

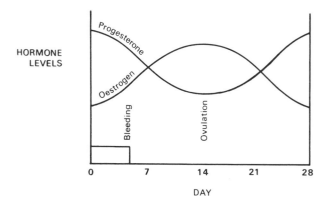

One factor which may affect the performance of women is menstruation. For some women athletes success follows almost exactly the pattern of their monthly menstrual cycle. To avoid losing a gold medal because the championships are a week too early (or late) many take drugs like the 'pill' to control their cycles. Similarly, there is evidence of gymnasts artificially delaying the start of puberty. This not only postpones menstruation, but prevents the laying down of fatty tissues.

3 What to do with Data

More Measurements

We have already, in Chapter 2, measured body features such as height, temperature and centre of gravity. In the scientific study of sport many more types of information are needed, each measured in a particular way. Each measurement, too, has its own unit. For example, speed is normally stated in metres per second. To work it out we must measure both the distance covered and the time taken.

S.I. Units

The English-speaking world has become increasingly used to the 'metric system' of measurement. Scientists have used metric units for a long time; they, and this book, use the international system of units known as Système Internationale (S.I. for short).

Table 3.1 gives the common S.I. units and their abbreviations.

Measurement	SI unit	abbreviation
length	metre	m
mass	kilogram	kg
time	second	s
speed	metre per second	m/s or ms^{-1}
acceleration	metre per second per second	m/s^2 or $m\ s^{-2}$
force	newton	N
work, energy	joule	J
power	watt	W

Table 3.1

Apart from time with its minutes and hours, all the S.I. units have the same sets of larger and smaller relatives. For example, a kilometre (km) is 1000 metres, a kilojoule (kJ) is 1000 J; a millimetre (mm) is 1/1000 m, a milliwatt (mW) is 1/1000 W. Less common are centi (1/100, e.g. cm) and mega (1 million, e.g. MW).

Results Tables

All measurements need to be recorded, and the simplest way to do this is to draw up a results table. This keeps the information well organised, tidy and easy to read. Some examples are shown.

Distance, m	0	10	20	30	40	50	60	70	80	90	100
Time, s	0	1.9	3.7	5.1	6.4	7.6	8.7	9.8	10.9	12.0	13.1

Table 3.2 Intervals in 100m sprint

Attempt	1	2	3	4	5	6	7	8	9	10
Score	7	6	7	6	5	6	5	3	4	3

Table 3.3 Practice effect

Person	Height, m	Mass, kg	Surface area, m^2	Body temp. °C
Adam	1.80	76	1.94	36.8
Brenda	1.45	50	1.38	37.1
Catharine	1.62	48	1.50	37.0
Dick	1.53	63	1.59	37.0
Esme	1.66	66	1.73	36.6

Table 3.4 Body Data

Before you record any results, make a table which has in it everything except the final results. This will save time, and make sure that all the data is recorded—and in the right place.

Person	Body weight, N	Step-ups Height, m	No.	Work, J	Time, s	Power, W
Frances		0.5	100			
Graham		0.5	100			
Helen		0.5	100			

Table 3.5

Bar Charts, Histograms and Graphs

Although results tables carry much useful information, data is often more easily understood if it can be shown in a more pictorial way.

BAR CHART   HISTOGRAM   GRAPH

Bar charts, histograms and graphs are ways of doing this, particularly if we are interested in the relationship between two things. They all have features in common, but there are also important differences.

Each has two axes, lines at right angles, which form the limits. The vertical axis is always marked with evenly spaced numbers which often, but not always, start from 0.

Bar Charts

The horizontal axis of a bar chart (Fig. 3.1) is not usually marked with numbers, but with words. Bar charts are used to show features which are distinct from each other, such as males and females, different sports or nationalities. For this reason, there is no definite order for the bars, and spaces are often left between them.

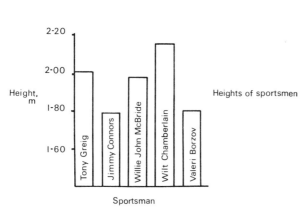

Fig. 3.1

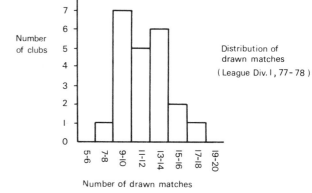

Fig. 3.2

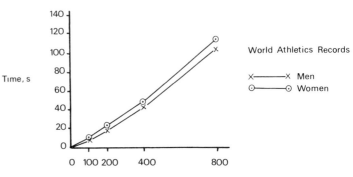

Fig. 3.3

Histograms

Histograms (Fig. 3.2) deal with continuous, numerical data: numbers in order. Often, this data comes in groups covering a range of values.

Graphs

In a graph (Fig. 3.3) both axes are marked with equally spaced numbers. Definite points (dots or crosses) are joined together. Graphs can be used to show any data in which both parts are sets of single numbers.

Selectagraph

You can use Fig. 3.4 to help you decide whether to use bar chart, histogram or graph.

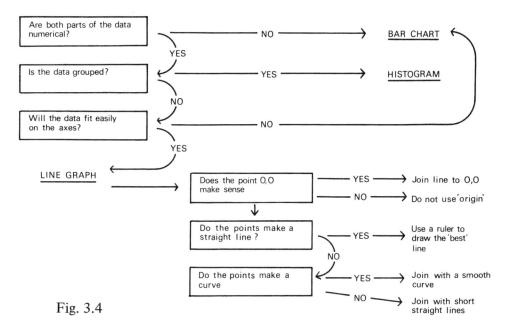

Fig. 3.4

Sports Records

Much useful data can be obtained from sports records—the Olympic Games, Football League tables, and so on. Presented graphically, features may become clear which were not so obvious in even the best of tables.

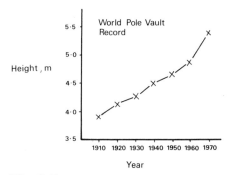

Fig. 3.5

Year	1910	1920	1930	1940	1950	1960	1970
Height, m	3.9	4.2	4.3	4.5	4.6	4.8	5.3

Table 3.6 Pole Vault Record

Take, for example, the world pole vault record:

It is easy to see from the graph that something dramatic happened between 1960 and 1970. Can you suggest what this might have been?

In fact, it was the introduction of the glass fibre pole.

Month	Aug	Sept	Oct	Nov	Dec	Jan	Feb	Mar	Apr
Points	4	11	8	6	5	5	2	5	11

Table 3.7 League points each month

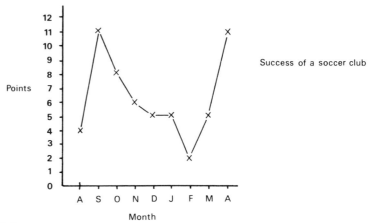

Fig. 3.6

Similarly, the level of success of a soccer club during the season can be seen more easily on a graph.

More examples will be found in Chapter 9.

Accuracy and Errors

Suppose everyone in the class has a stop-watch, and times the same event. Would you expect them to record exactly the same time?

No. There would be several reasons for errors, some faster, some slower. See how many you can think of.

You probably thought of reasons like these:

(a) errors in the watches themselves;
(b) human error—too early or too late in starting or stopping the watch;
(c) viewing error—not everybody could be exactly in line with the finish;
(d) inaccuracy in reading the watch, especially if between two markings.

In fact, there are errors in the best scientific measurements. The main point is to know that they exist, and to try to avoid them, or to make allowances for them. Let's look at a few examples.

At a normal athletics meeting—we'll come to the Olympics later—the timekeepers stand or sit alongside the finishing line. It didn't take long to find out that the times for 400m races and longer were much more accurate than those for the sprints. There were two reasons for this. Look at the track diagram, and try to think of them.

One reason is purely mathematical, 100m takes about 10s; 10000m takes about half an hour. An error of 0.1s makes a lot of difference in 10s, very little in 1800s! The longer the race, the less important

Disagreement or Error?

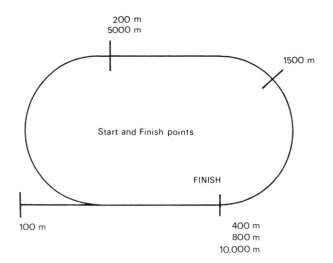

Electronic Watch

a very small error. To be as inaccurate as 0.1s in the 100m would mean an error of nearly 1½ minutes in the marathon!

The second reason is more scientific. When starting pistols were first used (before then it was flags), watches were started as soon as the shot was *heard*; now they are started when smoke from the gun is *seen*. The speed of sound in air is about 300m/s; the speed of light is 300000000 m/s. Sound takes 0.3s to travel 100m; light takes virtually no time at all. To make things worse, it is the shortest races which start furthest away from the timekeepers.

Modern electronic watches are extremely accurate, but this is no help if the timekeeper doesn't start and stop it at the exact instant. Electronic timing as used in all major athletics and swimming events removes the human timekeeper altogether. The clock starts automatically as the gun is fired, and stops as the finishing line is crossed.

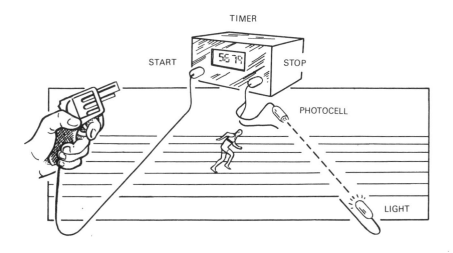

Measuring the High Jump

Measuring distances, too, can suffer from both mechanical and human error. The height of a jump will be inaccurate if the bar or ruler are bent, or if the judge is not tall enough to see clearly, or even if the ground is uneven. Again, modern technology has removed both kinds of error, at least in important competitions.

Two Kinds of Measurement

Disputed goal wins final

Disqualified after finishing first

CROWD BOOS JUDGES

No-jump costs title

WORLD RECORD NOT RATIFIED

WINNING GOAL IN INJURY TIME

Run out on last ball

These kinds of headline remind us that there are two types of measurement:
  objective—a numerical value which can be measured by instruments or counted;
  subjective—a score based mainly on the views of the judge, whatever the guidelines may be.

Let's look at some examples; try to decide whether each 'measurement' is objective or subjective.

(a) a long jump of 7.95m
(b) football result, 3–2
(c) figure-skating score, 7.8
(d) a round of golf in 72 strokes
(e) a boxer winning on 'points'
(f) a gymnast scoring 9.7
(g) a swim taking 53.8s
(h) a snooker break of 84

In some sports, objective measurement is straightforward, counting strokes or goals. In others, the measurement is objective, given the problems already discussed: times and distances. A third group of sports has a largely subjective scoring system. Some of these, such as figure skating, even give points for 'artistic impression'. In these sports, no single judge's score is 'correct', and an average is taken.

Look through the list of sports on page 12, and find others which have some subjective measurement.

Even in the most objective sports, though, there are subjective decisions to make. Referees, umpires and judges have only a fraction of a second

to decide whether a ball or toe was over a line, or whether time was 'up', or whether a player was offside, or . . . The measurement may be objective, but it is a subjective decision as to whether it should be made at all.

What Does it Mean?

Recording figures in tables and drawing graphs is not the end of this business of data handling. We must always try to draw some conclusions; we have to interpret the results, to try and explain them.

Have another look at Fig. 3.6 on page 32. The club obviously played badly in the winter, particularly in February. Or did it? There are several other possible explanations; a few are given here:

(a) fewer matches played (bad weather etc.);
(b) more difficult opposition;
(c) fewer home games;
(d) several key players injured.

In fact, there is no information to tell us how well they played at any time, only their success in getting league points.

Let's look at some more results and see what they tell us. Fig. 3.7 shows the numbers of sportsmen treated for brain damage at one unit in a year. What do the figures tell us?

A first idea is that soccer is very dangerous, while boxing is relatively safe. A moment's thought tells us that this cannot be so. Far more people play soccer than compete in boxing; we need to know how

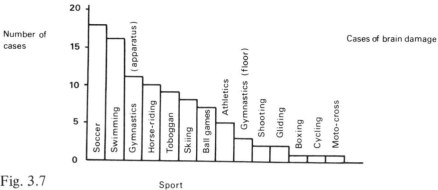

Fig. 3.7

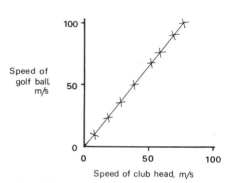

Fig. 3.8

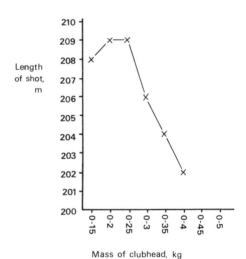

Fig. 3.9

many before we can reach any valid conclusions.

Bar charts normally give little information about reasons. Histograms and graphs are much better. For example, a straight line graph shows a direct relationship between the two factors: double one, and the other is doubled (or halved).

A curved graph may have one of several meanings. Often it suggests that a limit is being reached, or that a third factor is more important.

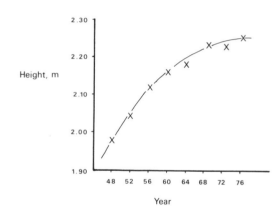

Fig. 3.10

In golf, it is the speed at which the club hits the ball which is most important (Fig. 3.8), so other factors produce curves like Fig. 3.9. In Fig. 3.10 a limit is being reached. Only a technical change, like a new vaulting pole (Fig. 3.5) will change the shape of the graph.

Questions

1. What are the SI units of (a) length, (b) force, (c) energy?
2. How many (a) millimetres in a metre, (b) Joules in a kilojoule, (c) watts in a kilowatt?
3. Draw a graph of the data in Table 3.2 on page 28.
4. From Table 3.2, which 10m was (a) fastest, (b) slowest?
5. Would you use a barchart, histogram or graph to show each of these?
   (a) Olympic marathon times, 1908-80
   (b) Number of medals won by each country
   (c) Numbers of batsmen scoring 0–49, 50–99, 100–149, 150–199, 200 + in a Test Match

(d) Number of passes before each goal in a hockey match
(e) Size of crowd at a football ground each Saturday
(f) Time of each runner in a race
(g) Amount lifted compared with body mass of each weightlifter.

6 What are the main reasons for error in
(a) timing a 100m race?
(b) measuring a discus throw?
(c) judging gymnastics?

7 Would more accurate timing at (a) the start, (b) the finish make a race seem faster or slower? Why?

Month	Aug	Sept	Oct	Nov	Dec	Jan	Feb	Mar	April
Points	8	5	5	3	2	0	0	2	7

Table 3.8

8 Table 3.8 gives the number of points gained by the same club as in Table 3.7 but in the following season.
(a) On one graph, draw data from both tables.
(b) What can you tell about the club's successes in the two seasons?

Extra: A Closer Look at Records

In Chapter 2 Extra we looked at some of the sex differences in running and swimming records. Here's some more information about them.

	Distances m	Men min	Men s	Women min	Women s	Men as % of women
Running	100		9.95		10.88	91.5
	200		19.83		21.71	91.3
	400		43.86		48.60	90.2
	800	1	42.4	1	54.9	89.1
	1,500	3	32.1	3	56.0	89.9
Swimming—	100		49.44		55.41	89.2
Freestyle	200	1	49.83	1	58.43	92.7
	400	3	51.41	4	06.28	94.0
	800	7	56.49	8	24.62	94.4
	1,500	15	02.40	16	04.49	93.6
Breast-	100	1	02.86	1	10.31	89.4
stroke	200	2	15.11	2	28.36	91.1
Butterfly	100		54.18	1	01.51	88.1
	200	1	59.23	2	07.01	93.9
Backstroke	100		55.49	1	01.51	90.2
	200	1	59.19	2	11.93	90.3
Medley	200	2	03.29	2	14.07	92.0
	400	4	20.05	4	40.83	92.6
Running, all distances mean						90.4
Swimming, all distances and stroke mean						91.7

The consistency is quite surprising. The range is only from 88.1% (100m butterfly) to 94.4% (800m freestyle). Overall, men are about 11% faster than women, regardless of distance, swimming stroke, or even when we compare running and swimming.

This kind of agreement is always worth looking for in any kind of data. We are looking for 'significance' in the figures: do they form some sort of pattern, or are they haphazard or random? In some cases, we may need a statistical test to check any apparent agreement.

Logarithmic Graphs

Look again at the table at the start of this extra section. Try to plot a graph of the running records, men's against women's. Assuming that you remember to convert all the times into seconds, the numbers range from 9 to 213, and from 10 to 236. Can you get them conveniently on to one axis? (The swimming times are even more spread, from 49 up to 6303: try getting these on the same axis.)

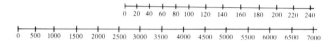

If we had included the 5000m, 10000m and marathon, the task would be impossible. The solution is to use a 'logarithmic' graph. There are three ways of doing this

(a) use special logarithmic graph paper;
(b) look up logarithms in tables, and plot these instead of the original numbers;
(c) mark the axis so that each position is ten times the value of the previous one:

Some examples of the use of logarithmic graphs are shown below. See if you can find others elsewhere in the book.

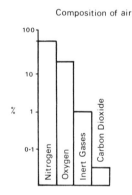

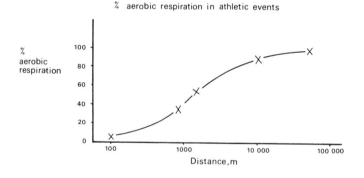

4 Analysing Movement

If we are to study any sport, we must have a way of looking closely at movements. Most movement in sport happens so quickly that the unaided eye is unable to appreciate it. This chapter shows a few of the ways of analysing movement, breaking it down into small units for study.

Flicker-book

Most of us have at some time drawn pin-men in the corners of the pages of a book. When flicked quickly, they give the impression of movement. You can make your own flicker-book by drawing on several pieces of paper, then stapling them together. Try your own sporting action, or copy the drawing.

You'll find it more effective if you

(a) use a simple action, and
(b) have very small stages between each drawing.

Using a flicker-book is really cheating, though. You either have to keep changing your drawings until they look 'right', or else you have to use frames from films—or someone else's drawings. Let's see how the drawings were obtained.

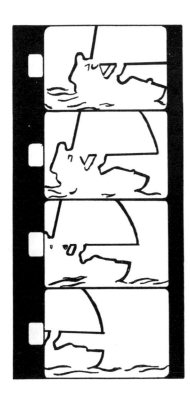

Film/Video-tape

Cine film is really a kind of transparent flicker-book. It is made up of thousands of separate photographs, taken at the rate of 18 or 24 per second. Looked at in quick succession they give the appearance of a continuously moving picture. Video-tape (TV film) is more continuous, but it still depends on the speed of scan.

The individual photographs ('frames') can be projected and copied, or they can be traced. For fairly slow actions, every frame is not needed. For very fast actions the camera has to run at a high speed, so that more frames are taken in the same time.

It isn't necessary, though, to make drawings. Running the projector at slower speeds, or frame by frame, is often enough for us to see how a movement is really made.

Still Photographs

A film, then, is a series of still photographs taken in rapid succession. It is possible to take a set of photographs during a sporting action with an ordinary still camera—as long as the film can be wound on quickly enough. Some modern (and expensive) cameras now have a motor-driven film wind. But even with a simple camera there is a way of making such a series:

1 Set up a simple sporting action, e.g. sprint start or tennis serve.
2 Take a photograph at the start of the action.
3 Wind on the film.
4 Repeat the action; take a photograph just after the start.
5 Keep repeating the action, taking each photograph slightly later than the previous one.
6 Mount the prints as if they were from a single sequence.

For best results, the action must be identical each time, and the background must stay the same. (Otherwise people may appear and disappear in what is supposed to be a fraction of a second!)

Stroboscope

The stroboscope has long been a scientific tool. Now it is increasingly common in entertainment ('disco' lighting) and car maintenance ('timing' light). There are two types of stroboscope:

(a) a regularly flashing light, usually xenon filled, which can be set to flash at any speed,
(b) a rotating disc, with regular slits, in front of a bright light.

When the light flashes on, any object in its path can be seen. When it is off, nothing is visible.

Any continuous movement will be seen at intervals. A ball in flight will appear, disappear, appear again.... Legs will seem to jerk. A bouncing ball may seem never to move at all.

The exact effect depends on how the speed of flashing compares with the speed of movement. For a regularly repeating action—running on the spot, squat thrusts etc.—the movement will appear as in the table:

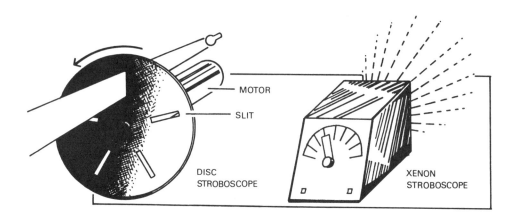

Flashing speed (relative to action)	Movement appears
very much faster	'normal'
slightly faster	slow motion
exactly the same	stationary
slightly slower	slow backwards!
very much slower	jerky

A camera with its shutter held open will record only the positions when the light flashes on:

Muscles and Movement

By using any—or all—of these techniques we can record and analyse movement. Any kind of movement: jumping, running; swimming; gymnastics; the flight of balls, shuttlecocks, javelins and disci; and many others. What sports have in common is the movement of human muscle.

The main muscles of the body were drawn in Fig. 2.2. One of their important features is that they work in pairs. Muscles can only do work when they contract, when they shorten. To move a joint therefore needs two muscles; one to straighten, one to bend. The best known of these 'antagonistic' pairs is the biceps and triceps of the upper arm.

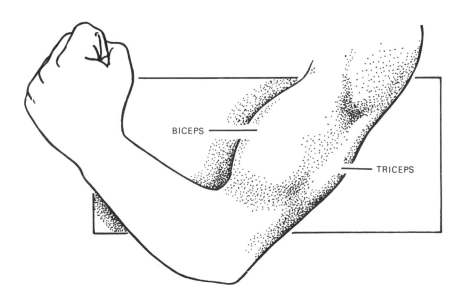

Muscles and limbs work like levers. Different joints are organised in different ways, but all are made up of

(a) pivot or fulcrum (the joint itself),
(b) load (the limb itself plus any object moved),
(c) effort (the muscle producing the force).

1st order levers are the most efficient, but are rare in human limbs. An example is the ankle.

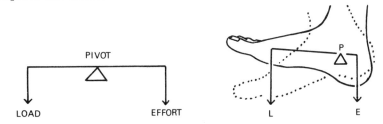

The foot also provides a 2nd order lever: standing on tip-toe.

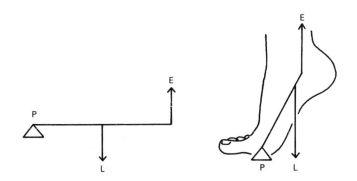

3rd order levers are the most common in the body. Lifting the forearm is an example.

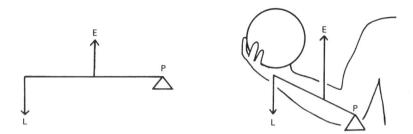

Movement of the body, or even of a single limb, is much more complicated than this suggests. The examples give much more information about the co-ordination of muscular activity in sport.

Close inspection of the drawing suggests that running can be divided into two stages:

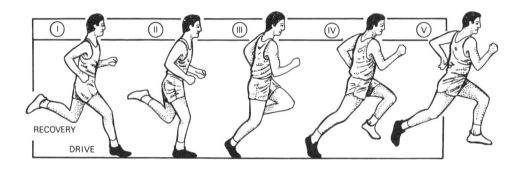

(a) drive, when the foot is in contact with the ground, and
(b) recovery, when the leg swings forward.

Drive. The foot lands, on the outside first (i), then takes the full weight of the body (ii). The knee 'gives' to avoid damage (iii), and, as the foot pushes backwards, the body moves forward (iv) and takes off (v).

Recovery. As soon as the toes leave the ground (iv), the leg is bent at the hip, knee and ankle (v). It swings forward at about twice the speed of the body (i). The recovery leg is highest just as the front foot lands (ii).

Much other information can be obtained from analysis of slow motion film. For example, measurement of stride length has shown that Olympic sprinters have a stride of about 1.15 times their height. The women's 100m in Montreal was won with exactly 50 strides.

If you've ever tried running with your arms folded (or even behind your back!) you'll realise that arm movements are also important. The reason for the rapid arm movements of a sprinter is to balance out the body twist caused by the powerful leg action. At slower speeds there is time for the trunk to absorb this twist, so energy can be conserved by less use of the arms. Murray Halberg won the Olympic 5000m in 1960 with a withered arm by leading early to avoid a sprint finish.

You can use the drawings to investigate some of the limb movements in swimming.

Newton's Laws of Motion

Isaac Newton is perhaps best known for his work on light and gravity. He would, however, have understood much of the science involved in sport—even though he never played himself. Most important are his three laws of motion.

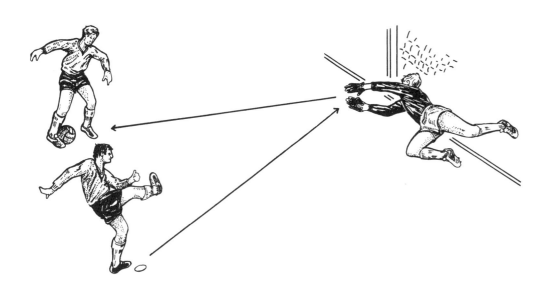

Consider a football on the penalty spot. A player kicks it towards the goal. The goalkeeper throws himself at it. The ball rebounds to a defender. He stops it.

Newton's first law is in action here. An object at rest stays at rest (ball on spot) until a force acts on it (kicked). Once moving, an object travels in a straight line (towards goal) until a force acts on it (goalkeeper). It then moves in a straight line again (rebound) until a further force brings it to rest (defender).

Would you rather be hit by a golf ball or a table-tennis ball? For the same strength of throw (the same force), a light ball accelerates faster than a heavy ball. But if they accelerate at the same rate, the heavy ball arrives with greater force.

In Newton's terms: acceleration = force ÷ mass
force = mass x acceleration

For example, a golf ball is 20 times heavier than a table-tennis ball, so falls with 20 times the force. (1 newton is the force which accelerates 1kg mass at 1 ms⁻².)

Have you ever watched a young child throw a ball? Often, he falls over backwards. It is easy to see that the child applies a force to the ball. Newton called this the 'action'. But the ball applies an opposite force to the child. This is the 'reaction'. Every force has an opposing, but equal, force.

In the diagram, the forces are in action/reaction pairs:

bodyweight (W) / contact force (N)
thrust (T) / friction (F)

Analysing Games

As well as analysing individual movements, it is often useful to do a sort of 'time-and-motion' study of a sportsman or of a match. This will give information about particular strengths and weaknesses, or about the performance of an individual team member. Here are two examples; they can easily be adapted to suit any sport.

1 Make an outline plan of a soccer pitch.
2 During a match, mark the movements of a player in a five-minute period. You can use different symbols to show walking, running, playing the ball.
3 Compare the same player at different times, or different players at the same time.

You can use this method to discover whether there is any truth in such commentators' statements as 'covered every blade of grass' and 'midfield players have a much higher work-rate'.

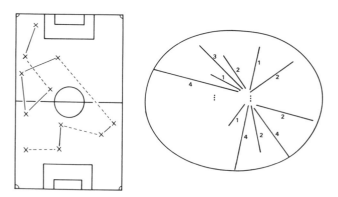

The second example concerns individual technique.

1 Draw an outline of a cricket field.
2 Each time the batsman plays a scoring shot, mark the direction and distance of the shot and the number of runs.
3 Use the results to compare scoring in front of and behind the wicket, leg- and off-side shots.

Energy

Motion or movement needs energy. Movement itself is a kind of energy, called motion or kinetic energy. Every moving object has some of its energy in this form.

But there are other kinds of energy. The table lists them, and gives a sporting example of each:

Type of Energy	Example
Kinetic	Movement of ball
Chemical	Food, fuel
Heat	Warming-up
Light	Floodlights
Sound	Bat on ball
Electricity	Electric scoreboard
Potential (gravity)	High jump bar
Nuclear (atomic)	Not in sport (yet!)

The cartoon strip shows a golfer from eating breakfast to the applause of the crowd. Can you identify the different kinds of energy involved?

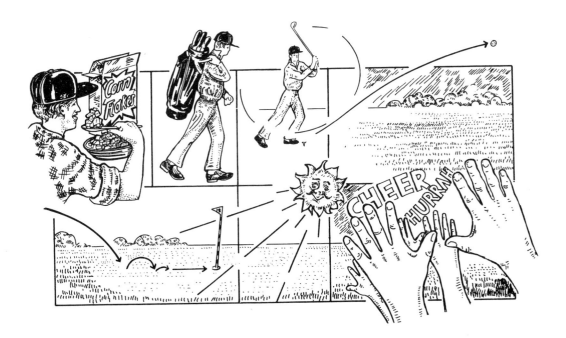

Questions

1. List four ways of analysing movement.
2. What are the two 'rules' for flicker-books?
3. At how many frames per second does cine film run?
4. Put these pictures into the correct order:

5. What are the two types of stroboscope?
6. (a) What are 'antagonistic' muscles?
   (b) Name a pair of such muscles.
7. What order are the lever systems shown?

8. (a) A sprinter is 1.8m tall. What is his probable stride length?
   (b) How tall would you expect the winner of the Montreal women's 100m to have been?

9 List six kinds of energy which occur in sport.
10 What kinds of energy are involved in
   (a) hitting a hockey ball?
   (b) a racing car?
   (c) electronically scored fencing?
11 Copy and complete these statements of Newton's laws of motion:
   (a) A resting object stays at rest unless.......................
       A moving object continues to move ....
       unless a force acts on it.
   (b) Acceleration, force and mass are related by the equation ......
   (c) Action and reaction are ....and ....
12 Look again at the cricket strokes on page 46.
   What would you advise the batsman?

Extra : Left and Right Hand

Three observations lead us to investigate the question of handedness in sport:

(a) more top-class tennis players are left-handed than would be expected (compared with the proportion of left-handers in the whole population);
(b) many left-arm bowlers (cricket) bat right-handed, but so do most right-arm bowlers;
(c) right-handed batsmen may be coached using the left hand only, never the right hand only.

What is the significance of these observations? Do we use the term 'left' and 'right' accurately?

Look at the tennis and cricket strokes shown. The arrangement of the hands is the same in each case. The tennis player is *left*-handed, playing the sort of backhand shot favoured by players such as Jimmy Connors. The cricketer is batting *right*-handed, playing a normal stroke. How can one be right-handed and the other left-handed?

Part of the difficulty is historical. For many years, left-handed people were thought to be 'odd'. Children were encouraged to write (and, presumably, to hit and throw) with the right hand. Some sports still reflect this: hockey players are forced to hold the stick in the right-handed style. Using a left-handed style and hitting with the back of the stick are forbidden.

If we look closely at a cricketer batting, we find that the top hand is used for power, the lower hand for guidance. We could argue that the 'left-hander' in fact bats right-handed, since it is the right hand on top which does most of the work. This is why single handed coaching concentrates on the 'wrong' hand.

A reason often put forward for the high proportion of left-handed tennis players is that being left-handed is a definite advantage. Because there are fewer of them opponents are not so used to dealing with them. Shots which a right-hander could not possibly get are fairly easy for a left-hander. Left-handers therefore win more games at first, and are more likely to reach club, county and national standard.

Momentum

Would you rather be hit on the head by a hockey ball or a table-tennis ball?
Would you rather be hit by a tennis ball lobbed gently at 5m/s, or one served at over 50m/s?
Why?
But why does it hurt less?

Scientists have a useful concept which they call 'momentum'. Momentum is a product of the mass and velocity (speed) of an object. To get its value, you simply multiply them together:

$$\text{momentum} = \text{mass} \times \text{velocity}$$

So, in the examples given above:

Ball	Mass, kg	Velocity, m/s	Momentum kg.m/s
Hockey	0.16	20	3.2
T-tennis	0.0024	20	0.048
Tennis	0.067	5	0.3
Tennis	0.067	50	3

We have already seen, in Newton's 2nd law of motion, that

$$\text{force} = \text{mass} \times \text{acceleration}.$$

And, as

$$\text{acceleration} = \text{velocity} \div \text{time},$$

we can relate all the quantities together. We find that

$$\text{force} = \text{momentum} \div \text{time}$$

This means that the force of impact when a ball hits something will depend on the momentum it has, and the time of contact. For a soft ball (or a soft surface), the contact time will be longer; the force will therefore be smaller.

It's not just balls which have momentum. People do, too. Provided, of course, that they are moving. There is an often controversial situation in rugby of the 'momentum' try.

'His own momentum carried him over the line,' says the commentator, or the referee, as a try is awarded. The laws of the game state that a try shall be scored 'if the momentum of a player, when held in possession of the ball, carries him into his opponents' in-goal and he first there ground the ball, even though it touched the ground in the field of play.'

As long as the player continues to move, the try is allowed. The actual amount of momentum depends, of course, on the mass of the players, since

$$\text{momentum} = \text{mass} \times \text{velocity}.$$

How much faster must a 60kg scrum-half run than a 90kg prop-forward to have the same momentum?

If the player stops, even for a moment, he has no momentum since his velocity will then be 0, and

$$\text{mass} \times 0 = 0$$

This is why a momentum try requires continuous movement.

Momentum can be passed on, from one ball to another, or from one player to another. A simple rule applies here too. The total amount of momentum is the same, before and after any collision.

5 A First Look at Ball Games

The Variety of Balls

The same basic scientific principles apply to all balls, but, we need to remember that there are a lot of differences in how balls are made. As well as size and mass, how a ball behaves is affected by what exactly it is made from, whether it is hollow or solid, and what kind of surface it has.

Simple data on size and mass of common balls is given in Table 5.1.

Ball	Mass, g	Diameter, cm
Cricket	160	6.4
Golf	46	4.1
Hockey	160	6.4
Rugby	410	19, 24
Soccer	420	22
Squash	24	4
Table-tennis	2.4	3.8
Tennis	57	6.4

Table 5.1

Bounce

One of the most important features of a ball is its amount of bounce. Each game requires a certain type of bounce. Imagine a football with the bounce of a golf ball, or playing table-tennis with a squash ball!

When a ball meets a surface, something has to 'give'. This may be the ball, the surface, or both. Usually, the ball is flattened to some extent, as shown in the photograph. The fact that the ball is 'elastic' means that it can store energy for a short time, and then release it as it springs back to its original shape.

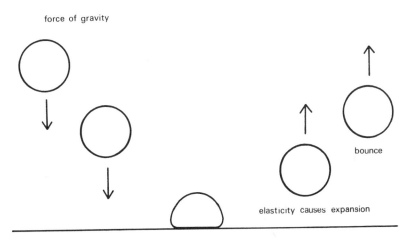

The bounciness of a ball is known by scientists as its 'coefficient of restitution'. Some idea of its value for a particular ball can be found quite simply.

1. Clamp a metre rule vertically.
2. Hold a ball with its bottom at the 1 metre mark.
3. Drop it.
4. Notice the height to which it bounces—look at the bottom of the ball again.
5. Repeat for other heights.
6. Draw a graph like Fig. 5.1.

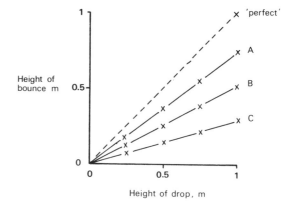

Fig. 5.1

A ball never bounces as high as it falls. This is because some of the energy used in squashing the ball is lost. This energy is converted into heat. For one bounce, it is a very small amount, but for repeated bounces it may become quite noticeable. In squash, this heat is very important as the hot ball bounces better than the cold ball. (See Chapter 6, Fig. 6.4, for more information.) Some of the energy may also be lost in squashing the ground. Some ball games have rules about the bounce of a ball. For example, a basketball dropped from 2m on to a hard floor must rebound to between 1.2 and 1.4m.

Surfaces

Why is the surface of each ball different?

The material of which a ball is made affects its surface. Leather panels used to make soccer balls can't be stitched to give the smooth surface of a squash or table-tennis ball. The way in which the ball is made can, however, have an effect. Machine-made cricket balls have an artificial 'seam' added to give some of the effects of the traditional, hand-made, type.

The most important factor in surface design is the effect which the surface has on the flight of the ball.

It is difficult to study the moving ball, so scientists use a wind-tunnel. Here, the ball stays still, and the air is moved past it. Smoke particles in the air mean that the flow can be clearly seen.

As a ball flies through the air, it disturbs the flow of air. Sometimes, the flow stays smooth, or streamlined:

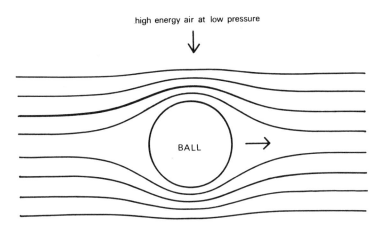

But, when friction between air and ball increases, the flow becomes turbulent:

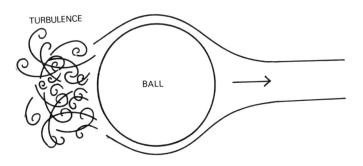

In turbulent flow, the air is disturbed and 'drag' occurs; this slows the ball's flight. The effect becomes greater as the speed of the ball increases. It also depends on the size and mass of the ball. A large light ball slows down more than a small heavy one.

A simple experiment will give an idea of this.

1. Take a selection of balls.
2. Find the mass and diameter of each; use Table 5.1 to help.
3. Throw each as far as you can. Use the same person, the same technique, and the same force for each ball.
4. Measure the length of each throw.
5. Don't forget your results table:

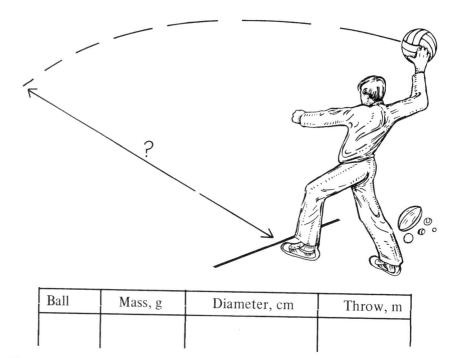

Ball	Mass, g	Diameter, cm	Throw, m

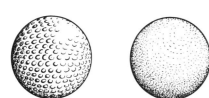

The surface of the ball does affect the amount of turbulence and drag. For example, the dimpled golf ball flies better than a smooth one.

Without air, in a vacuum, any ball would travel further. The graph (Fig. 5.2) shows the difference in range between cricket balls thrown in air and in a vacuum. Notice that the effect is greatest on the faster throws: the strong arm suffers more from air resistance than the weak arm.

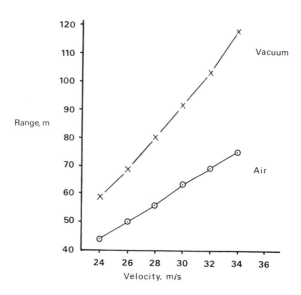

Fig. 5.2

Friction

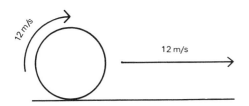

Friction is the force which opposes the movement of any two surfaces in contact. When a ball moves across a surface, there is bound to be friction between them. The amount of friction is greatest when the ball is sliding or skidding. It is least when the ball is rolling perfectly. Then, the speed of rotation is exactly the same as the speed of forward movement. If a resting ball is hit or kicked, it first skids, then begins to roll as it picks up speed.

By deliberately spinning the ball as it is hit, this skidding can be reduced, or even totally removed. This is normal practice in snooker and football.

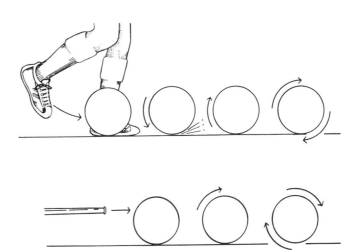

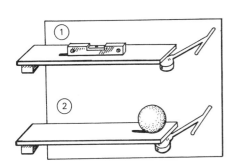

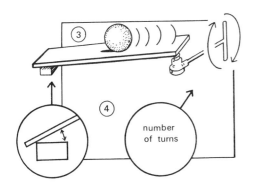

The frictional states of different balls can be compared by simple experiment.

1. Support a plank on a jack and a block so that it is perfectly level.
2. Put the ball on the plank at the jack end.
3. Carefully raise the jack until the ball rolls.
4. Measure the number of turns on the jack and/or the angle of the plank.
5. Repeat the measurements for other balls.

'Hitters'

It's strange how different words are used for the various hitters we use in ball games. Make sure you know which is which by matching these up:

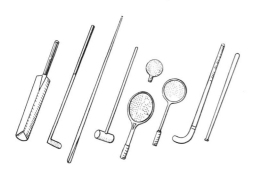

BAT	Badminton	Billiards
CLUB	Cricket	
CUE	Croquet	Golf Hockey
MALLET	Snooker	
RACKET	Squash	Tennis
STICK	Table-tennis	

Each type of hitter has its own features, but they all have one thing in common in the way they are used. Each is actually in contact with the ball for a very short period of time. The time of an actual hit is only a tiny fraction of a second. For a golf drive it may be only 0.0005s; in tennis, perhaps 0.005s. This very short contact time means that enormous forces have to be used to move the balls at their normal playing speeds, since

$$\text{force} = \text{mass} \times \text{velocity} \div \text{time}.$$

To get a ball speed of 70 m/s, the forces needed are 800 N in tennis, and over 6000 N in golf. This is equivalent to the weight of a small family car!

	Golf	Tennis
Ball mass, kg	0.046	0.057
Impact time, s	0.0005	0.005
Force, N	6440	798

It is possible to find out these forces for yourself.

1. Wrap a ball in aluminium foil, then find its mass.
2. Fix a piece of foil to the face of a bat or racket.
3. Connect wires from the foil on bat and ball to a timer. Wire it up so that the timer starts when contact is made, and stops when contact is broken.
4. Use a stop-watch (or photocell timing system) to time the flight of the ball across the room, and also measure the distance covered.
5. Hit the ball; record the data.

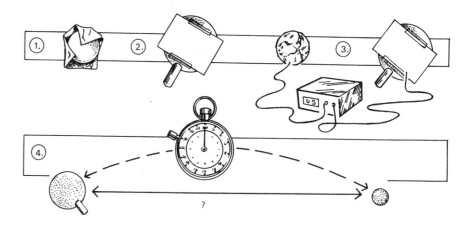

Work out the hitting force like this:

$$\text{force (N)} = \frac{\text{mass (kg)} \times \text{velocity (m/s)}}{\text{contact time (s)}} \text{ and velocity (m/s)} = \frac{\text{distance (m)}}{\text{flight time (s)}}$$

mass of ball (g)
÷
1000
×
distance (m)
÷
flight time (s)
÷
contact time (s)
=

The answer will be in newtons. Compare your figures with the examples given. Try it for different balls and hitters.

The material used to make a hitter will, as with a ball, affect the flight of the ball it hits. The small dents in a cricket bat tell us that energy is lost as the ball is hit. The strings of a tennis racket are very elastic, so little energy is lost. A racket with two sets of strings ('double strung') has been banned because it is too elastic. Friction between ball and hitter may also be important; this is why the snooker player chalks the tip of his cue.

Try to predict how unusual combinations of ball and hitter will behave. Then try them out. You might gain more understanding of some of the ideas in this chapter.

Questions

1. What five things may affect the behaviour of a ball?
2. What do scientists call the amount of 'bounciness'?
3. What are turbulence and drag?
4. Look at Fig. 5.2. What is the range of a throw of (a) 26m/s in a vacuum, (b) 32m/s in air?
5. What is friction?
6. What must be true of a ball if it is rolling perfectly?
7. What force is needed to give a squash ball a speed of 10m/s, if the impact time is 0.005s?
8. How can we tell that energy is lost in
   (a) the impact of cricket bat and ball?
   (b) the repeated hitting of a squash ball?
9. Look again at Fig. 5.1 on page 52.
   Suggest which balls might produce lines A, B and C.

6 More about Ball Games

The previous chapter introduced many basic ideas about the science of ball games. Here, we shall look in more detail at some of them, and other features of specific games.

Golf

Except for colour, one snooker ball is very much like another: the same mass, the same diameter, the same surface texture. This is not true with balls used in other sports. As we shall see, the temperature of a squash ball, the shininess of a cricket ball, the spin of a tennis ball are all important variables. The critical features of a golf ball are its composition and its surface.

Conventional golf balls are made of an extremely long rubber thread, wound very tightly around a heavy core. The higher the compression, the greater the elasticity, the faster the ball flies.

The newer type of ball is solid, made of plastic. It compresses less on impact, so it travels even faster. The plastic dimpled cover is not needed now to hold the thread in place, but it is not just to make it look nicer. The dimples increase the lift on the spinning ball, making it travel further.

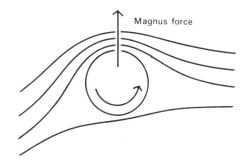

Magnus force

The spin of a ball has an enormous effect on its flight. The spin produces extra lift by the 'Magnus' effect. The pressure beneath the spinning ball is greater than the pressure above it, resulting in lift. There are some important factors which affect the amount of this spin-produced lift:

(a) it is increased by a rough surface, hence the importance of dimples;
(b) the amount of spin depends on the exact construction of the ball;
(c) it alters the 'best' angle of projection from 45° to 20°.

The relationship between angle of impact and the amount of spin-produced lift is interesting. A perfectly smooth ball projected at 45° with no spin at all would actually carry further than a 'normal' hit. The world's golf courses, however, are still waiting to see a golfer who can hit a perfect shot at 45° with a putter from a tee half a metre high!

A simple experiment may give some idea of the effect of dimples, and compare different kinds of dimples. As measuring flow through the usual fluid—air—is difficult, we make use of a rather denser fluid—water, or, better still, oil.

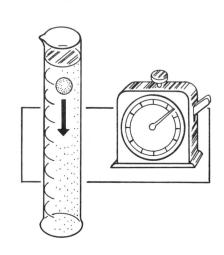

1 Find a smooth (old, worn) golf ball, and one with good dimples. (A third one with hexagonal dimples will give an extra comparison).
2 Fill a large measuring cylinder with liquid.
3 Hold one of the balls on the surface of the liquid.
4 Start a stop-clock as you release the ball.
5 Measure the time it takes to fall.
6 Repeat several times for each ball.
7 Tabulate your results, and work out the average rate for each ball.
8 Try spinning the ball to see whether this has any effect.

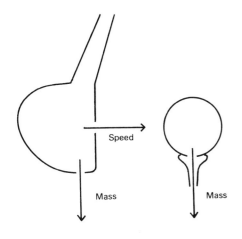

Is there any difference between the balls?

If you are interested in the effect of dimples, read the Extra section at the end of this chapter.

We have already seen (Chapter 5) that the speed of a golf ball depends on the force with which it is hit. This force depends largely on the speed at which the club is moving at the moment of impact.

There is a direct relationship between the two speeds which we can calculate. Let us assume that the ball has a regulation mass of 46g, and the clubhead a mass of 200g. Using the calculator:

> mass of club (200g)
> +
> mass of ball (46g)
> =
> M
> C
> mass of club (200g)
> ×
> 1.7
> =
> ÷
> MR
> =

This gives a result of 1.4: the ball will travel 1.4 times faster than did the club. You can vary the figures in the calculation to find out what would happen if you used a heavier (or lighter) club and/or a heavier (or lighter) ball. But, for a normal club and a regulation ball, the ball will travel at 1.4 times the speed of the club. So a manufacturer claiming a 40% increase in speed off the tee would actually be saying nothing at all about his ball. The rules of golf set the maximum ball mass at 46g. In the United States the ball must be at least 4.25cm in diameter. In Britain the minimum is 4.1cm.

The actual club used, of course, has much more effect than that of its mass on the speed of the ball. Fig. 6.1 gives an idea of the range which can be expected from various clubs, and the sort of flight path.

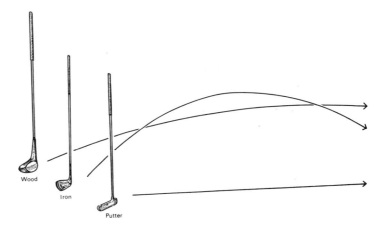

Fig. 6.1

Angle	Spin, rev/s
20°	57
30°	83
40°	106

Differences are brought about by the angle of the club face; this affects flight path, spin and speed. Examples are shown in Fig. 6.2.

As the angle of the club face increases, there is an increase in the angle of take-off. This means less speed, but more spin. More spin, as we have seen, means more lift.

Spin also has an effect on the bounce of the ball when it finally completes its flight through the air. A golf ball normally has back spin. The

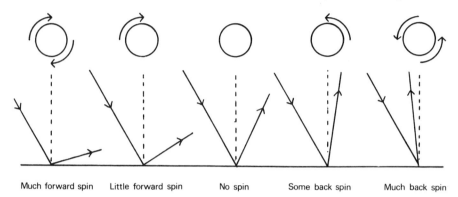

Fig. 6.3

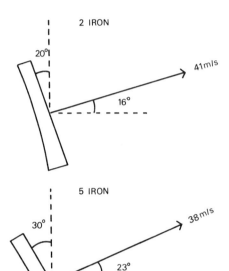

more spin, the more sharply it bounces. With enough spin, it may even bounce back on itself. In other games, the ball may have forward spin as it bounces. The effects of spin on bounce are shown in Fig. 6.3. The bounce of five balls landing at 30° is shown to depend on their spin. Forward spin = lower bounce. Back spin = higher bounce.

Tennis

Top Spin

can be given to the ball by hitting it with the racket moving upwards across the ball. This spins the ball in such a way that the flight path dips. A much higher landing angle means that the ball 'kicks' on bouncing, even though the spin makes the bounce angle slightly less than expected.

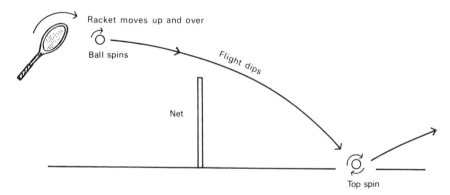

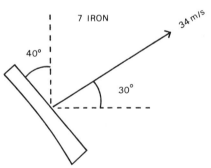

Fig. 6.2

Back Spin

is produced by 'chopping' underneath the ball. The flight path is more looped, but the ball bounces higher because of the spin. It may (Fig. 6.3) even bounce back on itself. This is the drop-shot.

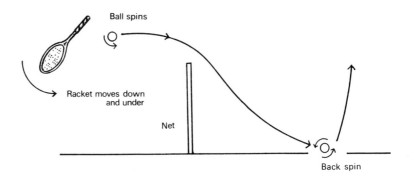

Cricket

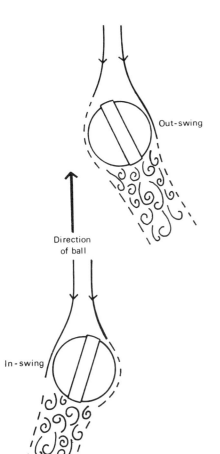

'Spin' and 'swing' are terms often used in cricket. The swing produced by a fast bowler with the new ball is due to a balance between streamlined and turbulent flow. (See page 53 if you've forgotten these already!)

If the ball travels with the seam at an angle, the air is faced with a smooth (seamless) side and a roughened (seamed) side. Flow will be streamlined on the smooth side, turbulent on the seam side. It is the different types of flow which make the ball swing in the air.

Swing can only occur if the flight is fast enough, and if the seam points in the same direction throughout the flight. It can be increased by roughening one side of the ball. (You should be able to work out which side.)

Spin bowling remains an art even when the science is understood. The spinning ball skids when it meets the pitch; the amount of sideways skid depends on the amount of spin, and on the length of time for which the skidding lasts. After the skid phase, the ball rolls in a straight line. The softer the wicket, the longer the contact, and the more skidding. Hence cricketing talk such as a 'sticky' wicket being a 'real turner'.

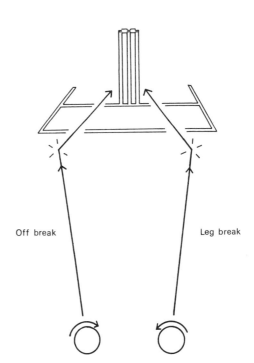

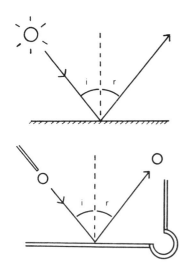

Angle of incidence (i) = Angle of reflection (r)

Snooker

The game of snooker takes its name, not from the 'potting of balls, but from the situation in which the path of the cue ball is blocked by another ball. An experienced player is able to swerve the cue ball in a curved path, so as to miss the obstacle.

The theory is similar to that of spin bowling: spin the cue ball so that it skids sideways, before rolling straight on.

Less skillful players make use of the cushion in such a situation. Here, two scientific principles are important. The importance of cueing at the right height (about 3.5 cm to avoid skidding) has already been discussed (Chapter 5). The law of reflection we tend to take rather for granted. Like light on a mirror, a ball will bounce off the cushion at the same angle as it arrived.

The most impressive part of professional snooker is not the potting of the balls, but the way in which the cue ball is directed ready for the next shot. To do this, players make use of top, bottom and side spins.

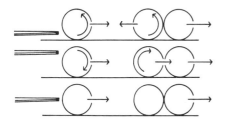

A cue ball struck low down will gain bottom spin. On hitting another ball it will roll back towards the cue. A ball given top spin by being struck above the 70% height will gain top spin. It will move forward faster after impact as this spin is turned into rolling. A 'stunned' ball hit dead centre will skid without rolling. It will stop 'dead' on impact, staying where it is. (This technique is also familiar to the shove ha'penny player!)

The cue ball may also be given 'side': side spin. This means that it is hit away from the vertical centre. After impact it will alter its direction of travel.

Squash

A squash ball is rather different to those met so far. Most important is its change of elasticity with temperature. As it warms up it becomes more and more elastic; its 'bounciness' increases. It bounces faster and further. This is why squash players need to 'warm up' the ball, as well as themselves. The graph (Fig. 6.4) shows the amount of variation.

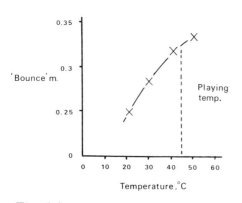

Fig. 6.4

Rugby

Most of the ideas discussed in these chapters can be applied to any kind of ball. The rugby ball is, however, rather different. The ovoid shape means that its flight is more complex. There are, though, a few simple investigations and explanations which are worth looking at.

Apart from problem areas such as lines-out and foul play (see Chapters

12 & 13), the most important changes in rugby in recent years have been:

(a) styles of goal-kicking;
(b) spin passes and throws;
(c) torpedo ('screw') kicks.

An apparently simple study of goal-kicking styles is full of difficulties. Unless we can build a goal-kicking machine, there are bound to be differences between kicks, even with the same players.

1 Select a suitable point on the pitch, e.g. exact centre, on the 22m line.
2 Take 10 kicks at goal, using the same style each time. Record the number of successes.
3 Repeat for other styles.
4 Repeat all the styles from different places. (Do each 10 times again!)

Does any style seem more efficient and accurate? Why have so many kickers adopted the 'round the corner' style? Are different techniques better for long kicks? Does the wind affect different styles in different ways? These are the kind of questions which the sport scientist can investigate, and try to explain.

At least with spin effects there's a definite answer: streamlining. The torpedo pass, throw-in and kick all try to do the same thing. They try to make sure that the ball travels 'point-first', giving the minimum of air resistance. This reduces the drag, and gives greater distance for the same impact force.

Questions

1. If a golf ball (mass 46g) is hit by a club (mass 200g) travelling at 30m/s, how fast will the ball travel?
2. From Fig. 6.1, which golf club
   (a) has the longest range?
   (b) gives the highest flight?
   (c) has its face at an angle of 30°?
3. Are the tennis strokes shown producing back-spin or top-spin?
4. Which law of physics is obeyed by balls hitting the cushion of a billiard table?
5. How is a conventional golf ball made?
6. What will happen to the cue balls shown after impact?
7. What is the effect of increased temperature on a squash ball?
8. Copy the two rugby balls, and draw the air flow around and behind them.

Extra: A Sport Science Advertisement

These extracts come from a 'scientific' advertisement for the 'Plus 6' golf ball.

'Many different tests have been carried out which we believe you will find completely meaningful and fair. Our new test has to be completely fair, because two balls are hit at the same time, instead of one at a time as in previous methods. This means that each hit is a true comparison—because all conditions, such as wind, rain, temperature, humidity etc., are clearly identical for each stroke.

Our test is true to life because the velocity of the club head is set within the range of human capability, and the angle of the head is 11°, that of a No. 1 wood.

Our green at the testing range is 175m from the hitting machine; in front of the green is rough.'

'We fired each (of four other brands) at the same time as the Plus 6 over a total of 100 hits. Throughout the test we alternated the Plus 6 between the right and left tee.

The results on the chart show how many of each brand reached the green:

Comparison ball		Plus 6	
Brand A	53	91	Test period
Brand B	11	94	23-27 June
Brand C	81	93	1975
Brand D	28	91	

These results speak for themselves.'

Do they?

There are clearly some very good features in the design of the test. You might like to list these. But it is also possible to make several criticisms: can you think of any?

Would you be persuaded by this advert to use the Plus 6 Ball?

'Plus 6' is manufactured by Uniroyal Ltd.

7 How the body reacts

Reflex save earns replay

NERVOUS SEMI-FINAL

Adrenaline flows, then champagne

Many of our ideas about how the body reacts during sport comes from statements like these. We read them in newspapers, hear them from commentators. Many of them are misleading; some are completely wrong. In this chapter, and in the next, we will look at what scientists call 'physiology'—how the body works.

Senses

We often speak of the 'five senses'.
Really, there are six:

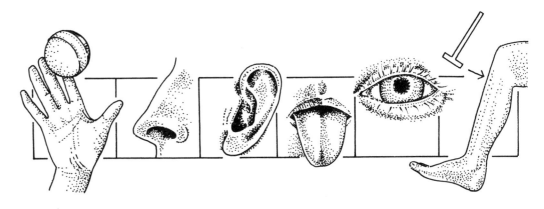

Of these, taste and smell are not of much use to the sportsman. Next come hearing and touch (including temperature and pain), with the most important being sight (vision) and 'muscle sense'. Let's start with this last one, as it is the least familiar.

Stretch Receptors

Inside muscles (and joints and tendons) are small sense organs called

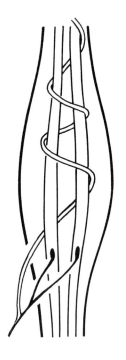

proprioceptors, or stretch receptors. These measure the stretch of muscles and tendons, the angles and pressures of joints. This is the basis of the system which keeps the body upright, and keeps a check on movements.

Little stretch
Little muscle tone

Muscles stretched
Tone maintains posture

Muscles very stretched
More tone needed

The more a proprioceptor is stretched—by whatever force—the more the muscle contracts. This contraction, which maintains body posture, is called muscle 'tone'.

Balance

The organs of balance (the semi-circular canals) lie within the inner ear. The three canals are set at right-angles to each other. This means that they can respond to movement in any direction.

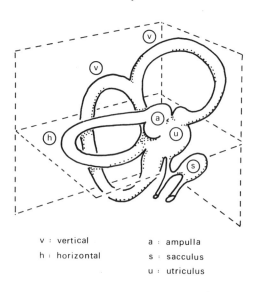

v : vertical
h : horizontal
a : ampulla
s : sacculus
u : utriculus

Sense organs in each canal are sensitive to the movements of the fluid inside. Without this system we wouldn't be able to position the body correctly while swimming, diving or even walking in the dark. (In the light we rely greatly on vision for balance.)

One problem is that the fluid keeps moving for a while after the head has stopped. The eyes and ears are then sending different information to the brain. Result: feeling giddy or dizzy.

Knee-jerk Reflex

Much body movement is controlled by reflexes. We are not conscious of everything our senses receive from outside (or inside). Reflexes mean that we can avoid danger (blinking, dropping hot objects) or adjust the body (digest food, cough, balance, change pupil size) without thinking. One of the best known of these is the knee-jerk reflex. If you hit the tendon, the leg kicks. Follow the route which the nerve impulses (signals) take:

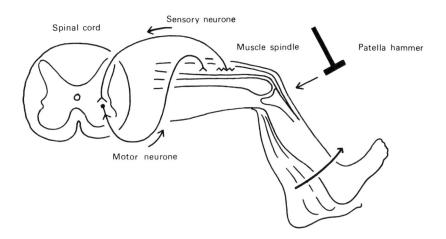

You can try this for yourself. It's quite easy to use this to measure the speed of body reflexes.

1 Connect 'hammer' (soft!), knee, heel and stool to a timer; wrap aluminium foil around them to make good electrical contact.

2 Cross one leg over the other, heel against stool.
3 Tap just below the knee. This will start the timer.
4 As the leg kicks, the timer will stop.
5 Take several measurements of reflex time, and find the average.

Reflexes

When a commentator talks of a goalkeeper making a 'reflex' save, is he using the word correctly? A true reflex is

(a) unconscious—no thought is needed,
(b) repeatable—the same response every time,
(c) universal—everyone reacts in the same way.

Most people's reaction to a ball speeding at their head is to get out of the way. Only a sportsman is likely to try to catch it—or to deliberately head it! It is often said, therefore, that a goalkeeper's action may be a 'conditioned' reflex. The classic example is of Pavlov's dogs which became conditioned to produce saliva on hearing a bell, rather than on smelling food. Now do you understand the drawing?

Others argue that sportsmen's responses have to be learned and practised.

Nerves

Sportsmen are often 'nervous' before events. Actors in the wings, students before exams and 'pregnant fathers' may all suffer in the same way. But when we are 'nervous' we are really being affected, not by our nerves, but by adrenaline. This is an important chemical messenger which will be discussed later (page 75).

Nerves actually change very little. Each nerve is made up of many nerve

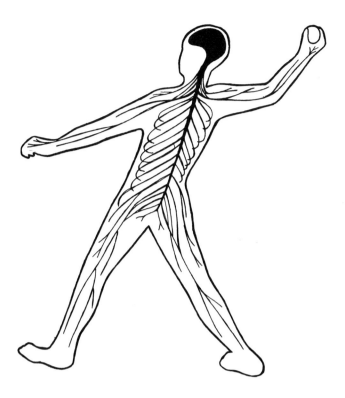

fibres, which themselves contain many individual nerve cells (neurones).

There are small chemical changes in the nerve as the electrical impulse passes along it. Other chemicals are used to cross the synapse—the gap between neurones. These changes take time; not very much, but a measurable amount. This is why even a reflex is not instantaneous.

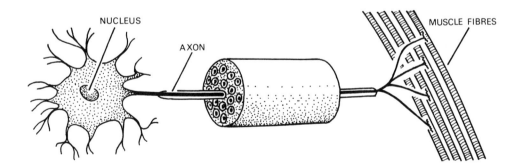

Reaction Time

True reflexes are the quickest way of turning a stimulus, such as a bright light, into a response—narrowing the pupils. They are not, however, what we are usually concerned with in sport.

The time which is important is the reaction time. This is the time between stimulus and response in a conscious action such as hitting a ball or jumping a hurdle. It is quite easy to measure using the equipment overleaf.

1 Connect the timer and switches.

2 Make sure the timer starts when switch 1 is on, and stops when switch 2 is pressed.
3 Set up the system so that the subject can't see switch 1 or the operator.
4 The operator starts the timer and switches on the light, by means of switch 1.
5 The subject presses switch 2 as soon as he sees the light. This stops the timer.
6 Record the time.
7 Reset timer, switch 1 and switch 2 (in that order).
8 Take several measurements, and work out the mean reaction time.

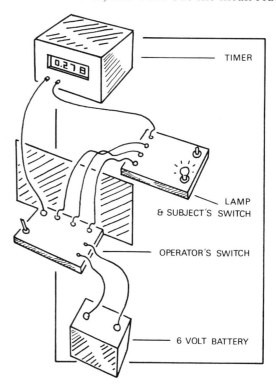

Results tend to follow the sort of pattern shown in Fig. 7.1: some improvement at first, but a definite limit. (It will of course make a difference where the subject rests his hand.)

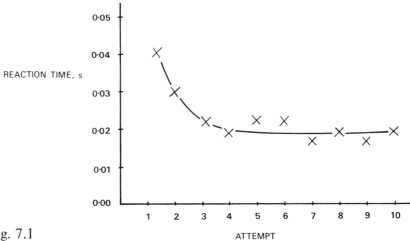

Fig. 7.1

Vision

Ask a sportsman which sense is the most important, and most will say sight or vision. It is certainly the one which is most well developed.

How do we see at all?
How do we see in stereo (depth)?
How do we see movement?
How do we see colour?

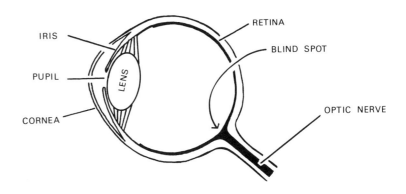

Our eyes have a complicated structure, but what they do is really quite simple:

light is bent by cornea and lens
as it passes through the pupil;
it is focussed on the retina,
where cells send impulses (messages)
to the brain, where we 'see' the light.

Seeing in Stereo

is very important in sport. Without binocular (two-eyed) vision we could not judge distance. England cricketer Colin Milburn and goalkeeper Gordon Banks both retired after losing the sight of one eye. You can get an idea of how important it is to have two eyes with this simple experiment.

1. Hold the book flat, just below eye level.
2. Close one eye.
3. Look along the dotted line in the margin.
4. Try to touch the ball.
5. Move the book to see how close you were.
6. Now try again, with both eyes open.
   If you can't do it now, see an optician!

It's more difficult to judge distance with one eye because the brain works out distances by comparing angles. It measures the angle of the light entering each eye. It's a similar method to finding your position with a map and compass: you need two compass bearings. One compass reading, like one eye, may give a fair idea, but it is never accurate.

Seeing Movement

clearly is another useful ability for a sportsman. We often speak of players 'having a good eye', and we use phrases like 'keep your eye on

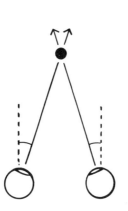
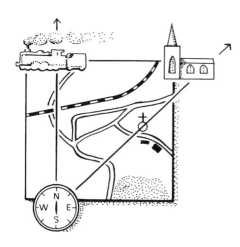

the ball'. These remind us that we are able to change the angle and focus of the eyes.

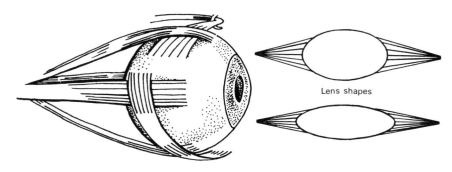

Lens shapes

This is done by various muscles. Some are fixed to the lens and change its shape. Others hold the whole eye-ball in position. These are able to move the eye in almost any direction.

Our eyes are best at seeing objects straight ahead. This is because there are more cells in the central part of the retina. When we see something 'out of the corner of the eye', we may only just be aware of it. Yet this ability is vital in sport, as it gives a chance to turn the eyes or the head for a clearer view. Scientists call this 'peripheral vision'.

Try this experiment on peripheral vision:

1. Get someone to look straight ahead.
2. Hold a coloured object behind him.
3. Move it slowly round, past his ear.
4. Record the positions of the object when he can
   (a) see it,
   (b) be sure what it is,
   (c) be certain of its colour.

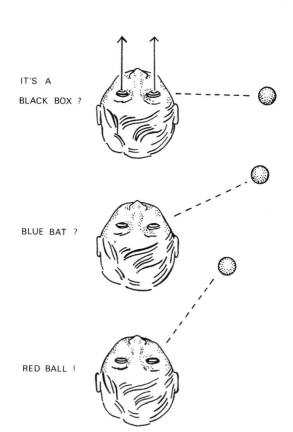

You may know that each eye has a 'blind spot'. This is the point at which the optic nerve leaves the eye, to carry impulses to the brain. At this point, there are no light sensitive cells, so any light landing here is not 'seen'. Use the cricketers opposite to discover your blind spot:

1. Close your right eye.
2. Hold the book at arm's length.
3. Look at the bowler, with your left eye.

4   You will be aware of the batsman, but do not look directly at him.
5   Slowly bring the book nearer.
6   The batsman will 'disappear' when his image falls on the blind spot of the left eye.
7   Keep bringing the book nearer.
8   The batsman will re-appear.
    You may be able to 'lose' the ball!
9   Try with the left eye shut to make the bowler vanish.

 - - - - - - - - - - - - - - - - - - - ● - - - - - - - - - - - - - - - - - - -

Pupils

are not actual structures. They are just 'holes' in the coloured iris. Each pupil is surrounded by two sets of muscles. One set makes it larger, so more light gets in. The other set makes it smaller, so that less light enters the eye. This change is a reflex, depending on the level of light. You can show this quite easily.

1   Notice the size of the pupils in 'normal' light.
2   Shine a bright light onto the eye.
    Notice what happens to the pupils.
3   Make the room dim by drawing blinds or switching off lights.
    Notice what happens to the pupils now.

In very bright light we tend to squint. This gives extra protection to the eyes until the pupils and retina have altered. Cricketers are always advised not to go in to bat straight from a dark pavilion.

Colour Vision

is important in only a few sports. Look again at the list on page
Which of these would be difficult for colour-blind sportsmen? Why?

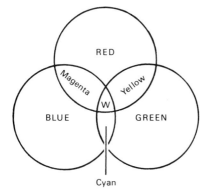
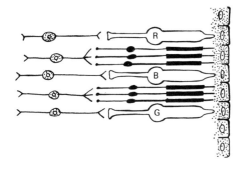

We see colour in a way similar to that in which a colour TV makes its colour. A colour TV has three tubes: red, green and blue. The retina of the eye has three kinds of cells for colour vision: red cones, green cones and blue cones. The fourth kind of cell—rods—work only in dim light, and in monochrome (black and white).

Hearing

It is difficult to assess the importance of hearing in sport. In team games, talking (and shouting!) are often valuable. For individual events, however, any sound may be disturbing. There is a balance between information and noise; like a badly-tuned radio, it may be important or distracting. (Engineers talk about the 'signal to noise ratio'.)

Try to think of some sounds which are (a) helpful, (b) distracting, or (c) have no effect.

As with vision, we hear in stereo; the brain compares the volume in each ear to work out the direction of sound. The ear itself, like the eye, is a complicated structure which does an apparently simple job:

> sound waves vibrate the ear-drum;
> the ear-bones (ossicles) amplify these vibrations and pass them on to fluid in the cochlea; receptor cells produce nerve impulses which the brain 'hears' as sound.

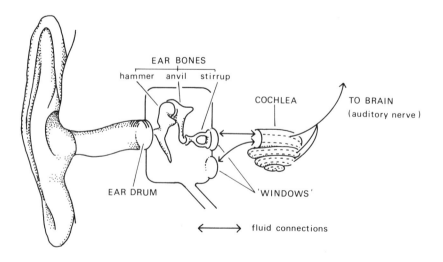

Hormones: Chemical 'Messengers'

Many of the body's activities go on for a long time. Nerves are not suited for the control of all of these. Instead, many reactions are controlled by chemical messengers—hormones.

The hormone most concerned with physical activity is adrenaline. Adrenaline is very similar to a substance produced by certain nerves. It has several very important effects:

> raises blood pressure
> increases heart rate
> makes more glucose available
> enlarges passages in lungs

Because of its production under stress, adrenaline is sometimes known as the hormone for 'flight, fight or fright'.

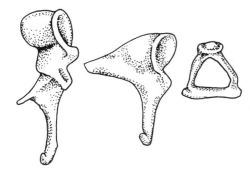

Questions

1. What is 'physiology'?
2. Name the six senses.
3. What are proprioceptors?
4. State three features of a reflex.
5. What makes us 'nervous'? What else does it do?
6. Look again at Fig. 7.1.
   What was (a) the first reaction time?
           (b) the fastest?
           (c) the slowest?
7. Even with both his arms, Lord Nelson would not have been as good a bowls player as Sir Francis Drake. Why not?
8. What are (a) the blind spot?
           (b) pupils?
           (c) the optic nerve?
9. What colours are 'seen' by the three kinds of cones?
10. Where would you find the bones shown?
    What are they called?
    What do they do?
11. What is done by semi-circular canals?
12. What are hormones? Name one. What does it do?

8 What is Fitness?

In general use, 'fitness' means being in good athletic condition, active and healthy. In sport science we need to be able to measure degrees of fitness, so we have to use a more precise definition.

There are really two different aspects of fitness. One is the ability to perform athletic tasks. We would not describe as fit someone who was unable to run upstairs. The second is the speed and ease of recovery from an activity. If, after running a short distance, we take ten minutes to catch our breath and for our heart to slow down, we can hardly be fit.

Basic Data

If we are to measure fitness, we must have knowledge of the resting state: heart rate, breathing rate and so on. We have already (Chapter 2) measured height, mass and temperature; now we will add some more data.

Find your pulse, on the thumb side of the wrist. Press with two fingers,

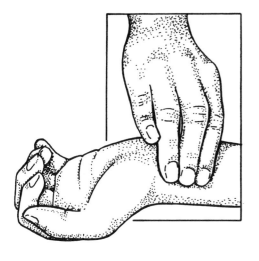

and count the number of beats in a minute. This is your pulse rate.

Count the number of breaths you take in a minute—in and out counts as one. This is your breathing rate. You may find it better to get someone else to count, as people often breathe differently when they are aware of it.

You may recall from Chapter 2 that there is often confusion over 'mass' and 'weight'. For work on fitness we need to use weight, in newtons. If you have platform scales which measure in newtons (N), use these to find your body weight. If not, you can get a rough value by multiplying your body mass (in kg) by 10.

body weight (N) = body mass (kg) x 10

Work, Energy and Power

Scientists define work in terms of force applied, and the distance the force moves:

    WORK = FORCE x DISTANCE
    joules   newtons   metres

So, if your 500N body jumps a height of 1m, you have done 500 x 1 = 500J of work.

Energy is a different way of looking at work: to do 500J of work needs 500J of energy.

Power is the rate of doing work. To do 100J of work in 1s takes much more power than doing it in 10s or 1 minute.

    POWER = WORK ÷ TIME
    watts   joules   seconds

Table 8.1 gives a few examples.

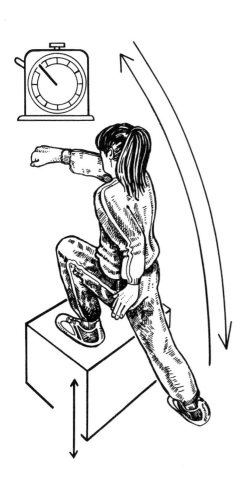

force, N	distance, m	work, J	time, s	power, W
500	2	1000	10	100
10	100	1000	50	20
100	6	600	60	
500	100		10	

Table 8.1

Step-ups, Pull-ups and Press-ups

Step-ups

are one of the simplest, and most useful, exercises in sport science.

1   Find a bench, box or similar about 0.5m high.
2   Measure its exact height, in metres.
3   Step up onto it, both feet, legs straightened. This is one step.

4 Use a stop-clock to find how long it takes you to do
   (a) 10 steps,
   (b) 50 steps,
   (c) 100 steps.
5 Draw out a results table like this, and fill in your own data:

A Body weight, N	B Step height, m	C No. steps	D Work, J (=AxBxC)	E Time, s	F Power, W (=D÷E)
		10			
		50			
		100			

6 Plot a graph of number of steps against power.

What kind of line do you get? What does this tell you?

Pull-ups

(or 'chins') concern a very different set of muscles.

1 Fix a beam, so that you can't reach it when standing on the floor.
2 Jump up and catch hold of the beam.
3 Get someone to start a stop-clock as you pull yourself up until your chin reaches the beam.
4 Measure the time you take for
   (a) 2,
   (b) 5,
   (c) as many as you can do.
5 During one of your pull-ups, get your partner to measure how far up your centre of gravity moves.
6 Produce a table like this, and fill in your own results:

A Body weight, N	B Height pulled, m	C No. pulls	D Work, J (=AxBxC)	E Time, s	F Power, W (=D÷E)
		2			
		5			

Again, you can plot a graph of pull-ups against power. Is the result similar to step-ups?

Press-ups

use another set of muscles. You can do ordinary press-ups, or, if you have parallel bars, 'dips'. Again, you need to measure how far your centre of gravity moves, and how many you do in a given time. The results can be recorded in the same way.

Effects of Exercise

While you do any of these activities—or any sport—you will notice several changes in your body:
- heart beats stronger and faster
- breathing quickens and deepens
- body temperature increases
- sweating
- aching muscles

All these 'physiological' changes have a cause and a purpose. In this section we will look at some of them in more detail.

Muscles

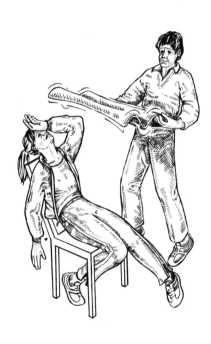

need energy. This energy comes from food, mainly converted to glucose (sugar) for use. To work most efficiently, muscles also need plenty of oxygen. Both glucose and oxygen are brought in the blood; wastes such as carbon dioxide are also carried away in the blood. This process of getting energy is called respiration.

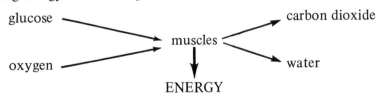

When muscles do extra work, more glucose and oxygen are needed. More blood must flow to the muscles. The heart beats faster, the blood vessels narrow to raise the pressure. More blood is sent to the muscles rather than other organs. You can see in Fig. 8.2 what happens to the blood flow as exercise becomes more strenuous.

Finally, however, it becomes impossible to get enough oxygen to the

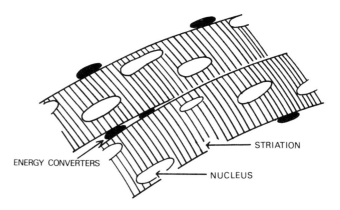

Fig. 8.1 Striated muscle (as in your limbs) showing 'energy converters' where glucose and oxygen are turned into energy for the muscles.

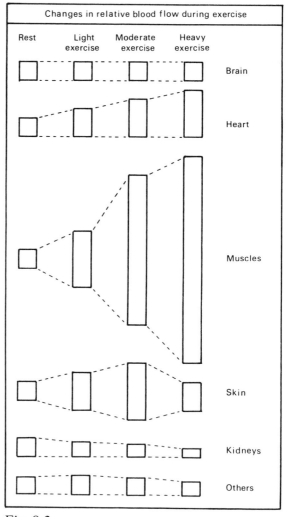

Fig. 8.2

muscles. They have then to use a different method of getting energy. Glucose is still used, but now the waste product is lactic acid:

Lactic acid is poisonous. After a while, there is enough of it to make the muscle ache. Eventually, it causes cramp, and the muscle will no longer contract. The athlete is forced to rest while the blood brings fresh supplies of oxygen.

Production of energy with oxygen is called aerobic respiration; production of energy without oxygen is called anaerobic respiration. There is further discussion of this, and of the 'oxygen debt', in Chapter 9.

Suffering from cramp

Sweating

Not all the energy produced in respiration is turned into muscle action. Some is turned into heat. The body can tolerate a small rise in temperature but, after a while, sweating begins. Sweat pours from the pores in the skin. Once on the surface, it evaporates. To evaporate, energy is needed; this energy comes from the heat of the body. As heat is lost, so the body temperature falls.

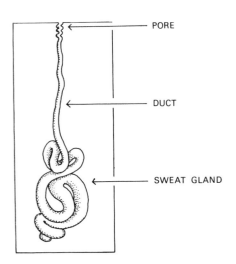

For every 1g of water evaporated, 2260 joules of energy are needed. This latent heat loss cools the body. The loss in sweat of water and salt, however, does cause problems; these are discussed in Chapter 9.

Heart Beat

starts in the pacemaker, but is controlled by nerves and hormones. During exercise, it is mainly adrenaline which produces changes in heart beat and blood pressure.

Some idea of fitness is given by the resting heart rate. An athlete's heart tends to be bigger and stronger than a non-athlete's; it can supply as much blood with fewer beats. So, the fitter you are, the lower your resting heart rate, although other factors such as age and body type have an effect. Average adult heart rate is about 72 beats per minute; an athlete's may be as low as 30.

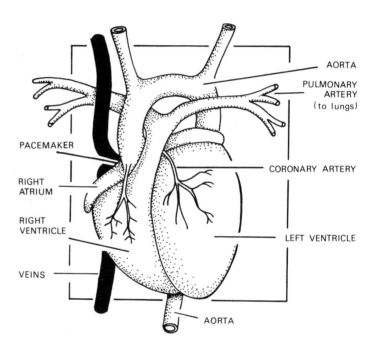

But the resting heart rate does not give enough information. We need to know what happens during exercise, and, just as important, how long it takes to return to normal. These values are the basis for an 'index of fitness'. One version of this is the 'Harvard Step Test', using the kind of step-ups described earlier.

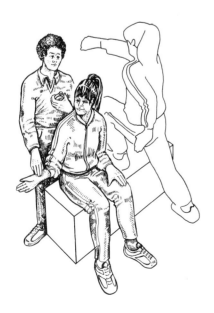

Harvard Step Test

1. Find a bench or box 0.5m high.
2. Step up and down, about every 2s for five minutes.
3. Rest for 1 minute.
4. Get someone to count your pulse for 30s.
5. After 30s pause, count pulse for 30s again.
6. Another 30s pause, then another 30s count.
7. Make a results chart like this, and work out your index of fitness.

```
*  Length of exercise    :           s
   Multiply by 100        :           (A)

   1st pulse count        :
   2nd pulse count        :
   3rd pulse count        : _____

   Add together           :
   Multiply by 2          :           (B)

   Divide A by B          :
```

This is your 'Index of Fitness'.
The higher the number, the fitter you are:

90+	80-89	70-79	60-69	50-59
superior	excellent	good	fair	poor

* Note: if you become exhausted, stop and record the exact time; use this in the calculation.

Breathing

is also greatly affected by exercise. However fast the heart beats, the blood can't carry enough oxygen if it isn't getting into the lungs. Two investigations will help measure the efficiency of your lungs.

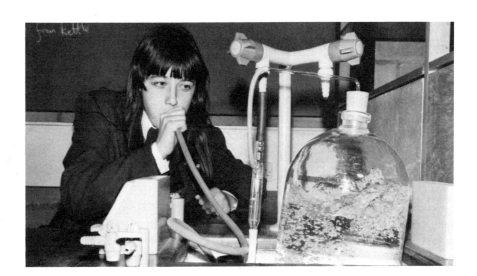

1. Set up the apparatus shown, with the clip open.
2. Turn on the tap, until the jar is full of water.
3. Turn off the tap; close the clip.
4. Breathe normally, blowing one of the breaths through the tubing.
5. Measure your 'tidal volume'.
6. Fill the jar again with water (repeat 1, 2, 3).
7. Take deep breaths, then blow out as much as you can through the tubing.
8. Measure your 'vital capacity'.

When we breathe in (inspiration) two sets of muscles are at work. The diaphragm is pulled down. The ribs are pulled upwards and outwards. This makes the chest expand, so lung pressure falls. Air is forced in from outside.

Two other sets of (antagonistic) muscles contract for breathing out (expiration). The diaphragm is pulled up and the ribs down and in. A smaller chest cavity means greater pressure, so air is forced out.

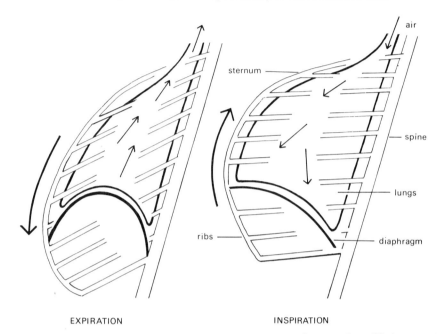

EXPIRATION   INSPIRATION

As well as how much air you can get into your lungs, the efficiency of breathing depends on how much oxygen you can remove from this air. The second investigation gives some idea of this.

1 Set up the apparatus shown.
   (To measure oxygen the liquid is alkaline pyrogallol. To measure carbon dioxide use potassium hydroxide solution).
   CARE: both liquids are corrosive

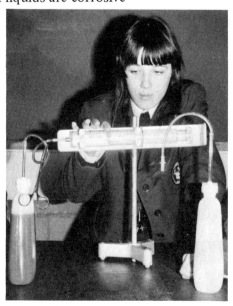

2. Fill both syringes with fresh air.
3. Fit the syringes in position, and open the clips.
4. Read the volumes (originally 100 cm³) now that some gas has been absorbed.
5. Repeat the measurements, using air which has been breathed out (exhaled or expired).
6. Compare your results with those in the table:

	fresh air	exhaled air
original volume, cm³	100	100
after CO_2 removal, cm³	100	96
volume CO_2, cm³	0	4
after oxygen removal, cm³	80	84
volume oxygen, cm³	20	16

The most important structures in oxygen uptake are the alveoli. These are tiny air-sacs, at the ends of the tubes (bronchioles) in the lungs. Each alveolus is surrounded by capillary blood vessels; oxygen crosses into the blood, carbon dioxide out.

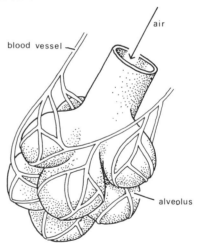

Bicycle Ergometer

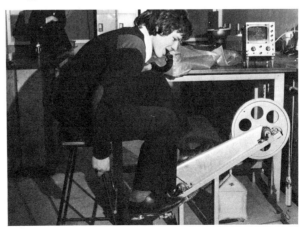

You have probably seen photographs or film like the one here, in which an athlete is having his fitness checked by a variety of instruments. The machine which is exhausting him is a bicycle ergometer. This lets him work against a measured force, so that the work done is easily compared with his oxygen consumption. The main features are shown in the drawing.

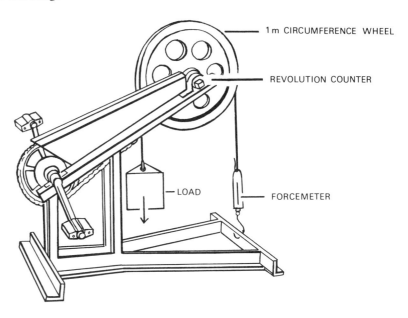

1 Note the reading on the forcemeter, at rest.
2 *either* count the number of revolutions in a given time.
   or   time a certain number of revolutions.
3 Take another forcemeter reading while pedalling.
4 Subtract the forcemeter readings; this gives the pedalling force.
5 Use a table like this for your results:

A Force, N	B No. revolutions	C Work, J (=A×B)	D Time, s	Power, W (=C÷D)

Food and Energy

We have seen that physical work requires energy, and we know that energy comes from food. Fig. 8.3 shows values for common foods and sporting activities.

Energy intake and output must be equal if the body's shape and condition are to stay the same. You may remember from Chapter 2 some of the strange diets on which sportsmen feed in order to maintain this balance.

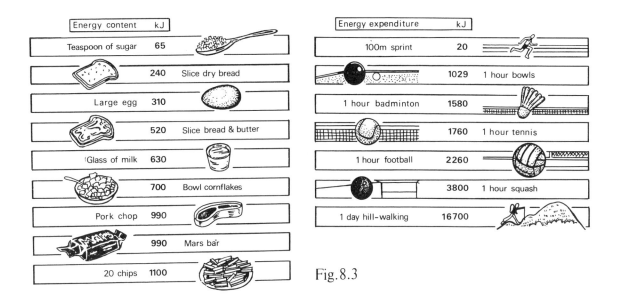

Fig. 8.3

Making Use of Food

In Chapter 7 adrenaline was introduced as the 'flight, fight or fright' hormone. In Chapter 9, hormones controlling body salt and water will appear. In this chapter are the hormones which control the use of food to produce energy.

One of the actions of adrenaline is to encourage the production of glucose. Liver and muscle cells store glucose as the starch-like glycogen. Adrenaline speeds up the conversion of glycogen to glucose which then passes into the blood. With the increases in blood pressure and heart beat, both food and oxygen can reach the muscles more quickly.

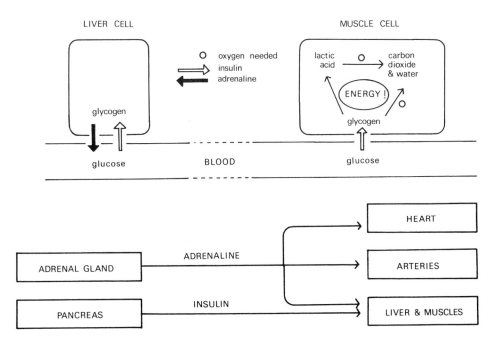

Another hormone—insulin—pushes glucose into muscle cells. There it can be used to produce energy, or converted into glycogen for future use. Its effect is to reduce the level of glucose in the blood. Diabetics have too little insulin, and cannot get enough glucose into the muscles. Activity for them may mean a dangerously high level of glucose in the blood. They can become sportsmen, but need to carefully control their diet and/or inject insulin regularly.

Questions

1. What two things do we mean by 'fit'?
2. What is (a) work, (b) power?
3. A weightlifter raises 750N a height of 2m.
   (a) How much work does he do?
   (b) If it takes 3s, how much power is needed?
4. Copy and complete this table, to show the differences between aerobic and anaerobic respiration:

	aerobic	anaerobic
Is oxygen needed?		
Waste products		

5. Look closely at Fig. 8.2 on page 80.
   Name one organ which, in exercise,
   (a) receives more blood,
   (b) receives less blood,
   (c) has a constant blood supply.
6. Calculate the index of fitness for a person with these results on a Harvard step test.
   length of exercise : 5 minutes
   pulse counts   : 84, 66, 50
   What rating would you give him?
7. What is shown in the drawing on the left? How is it used?
8. Use Fig. 8.2 to answer these questions:
   (a) Which contains more energy,
       a slice of bread or a glass of milk?
   (b) Which uses more energy,
       1 hour's squash or 1 hour's tennis?
   (c) How much tennis could 2 Mars bars support?
   (d) How many glasses of milk would be needed to replace the energy used up in a soccer match?
   (e) From the foods listed, work out a day's menu for a hill-walker.
9. Copy and complete Table 8.1.

Extra: Efficiency

Most machines are not very efficient. As much as 80% of the chemical energy in the fuel ends up as heat, only 20% as useful kinetic energy. Human machines (muscles) are better than most man-made devices, but they still convert less than half the energy of food into useful work. And, of course, some of this kinetic energy may be wasted by poor technique. Useful work is, however, more difficult to measure than total physical work.

Efficiency can be defined as

$$\text{efficiency} = \frac{\text{useful work done} \times 100}{\text{energy used}} \%.$$

We can measure the work done by means of step-ups, bicycle ergometer or other similar methods. But how can we measure the energy used up?

The rate at which the body uses energy is called the 'metabolic rate'. Even at rest this varies from person to person. Endomorphs tend to have low resting metabolic rates; ectomorphs high rates. We can get an idea of the metabolic rate by measuring the amount of oxygen used up. Earlier in the chapter we collected enough data to do this:

volume of air per breath (litres)
×
% oxygen removed
×
number of breaths per minute
÷
100

This will give the volume of oxygen used per minute. The amount of energy this produces depends exactly on the 'fuel' being used: glucose gives more energy per litre of oxygen than fat. The table shows how this varies:

Food	Energy, kJ/litre oxygen
glucose	21.1
mixed	20.1
fat	19.6

If we now multiply the amount of oxygen used per minute by this factor (19.6, 20.1 or 21.1), we will get a value for the energy used in a minute. For a mixed diet,

energy used = oxygen uptake × 20.1
kJ/min   litres/min

Efficiency is then given by

work done (kJ)
÷
length of activity (mins)
×
100
÷
energy used (kJ/min)

9 Athletics

Athletics, or track and field, is the simplest group of sports. The minimum of equipment is needed, and, apart from relays, competition is between individuals, even though their results may be added in some way. In this chapter we will look at several events, and at a 'special' meeting: the 1968 Olympic Games.

Sprinting

There are two vital features to the short sprint races: the start and respiration.

The start is more important in short races, as there is less time to make up any lost ground. Sprint starts as we know them were invented by fairground gamesters. They would boast that they could beat any opponent. So confident were they that they offered to start from a 'supine' position, lying on their backs. Try it for yourself:

1. Lie on the ground, face up.
2. Roll over onto your front.
3. Push yourself up on hands and feet as if to run.
4. You will find yourself in the sprint start position.
5. Compare starting like this with starting in a perfectly upright position, either against someone else, or by timing a short distance run.

The value of the sprint start is that it gives maximum forward thrust, and the lower running position offers less air resistance. More thrust and less drag means more speed.

A sprinter can run 100m without using much oxygen. During the 10s or so of the race he respires anaerobically (see page 81), and so builds up an 'oxygen debt'. The high level of lactic acid means that extra oxygen is needed after the race, as the debt is paid back. In longer races, the amount of anaerobic respiration is reduced so that large amounts of lactic acid do not build up. There is therefore a smaller oxygen debt.

Race	100m	800m	1500m	10000m	Marathon
% anaerobic respiration	95-100	65	45	10	2

Table 9.1

Hurdling

Watch a beginner hurdle, and you see two actions: running and jumping. Watch an expert, and you see a continuous, flowing movement. Ideally, the centre of gravity should rise only a little above the hurdle, in a smooth curve:

Important features of good hurdling are:
(a) Centre of gravity rises little; the high point is near the hurdle.
(b) The shorter time and longer distance spent off the ground, the better; this gives maximum speed.
(c) Wide separation of the legs, with fast backward and downward leg movements.
(d) No pause on landing.

Distance Running

The problems of the middle- and long-distance runner are rather different from those of the sprinter. Here, the emphasis is on stamina rather than power, on economy rather than strength. As more energy is needed, more stored food is used. The problem may not lie in getting the glucose and oxygen to the muscles, but in having enough glycogen (stored glucose) and fat in reserve.

Several methods have been devised to try to increase these reserves, many of them very strange. One technique is to first empty the reserves by eating a high protein, low carbohydrate diet during training. Then, just before the event, a 'saturation' diet of energy-rich foods is eaten. This is thought to produce more stored glycogen than the body can normally hold.

Some athletes have also tried to increase the oxygen capacity of the blood, to increase aerobic respiration further. One idea would be to take some blood from the athlete (like a blood donor) and store it. He would replace it naturally in a few weeks. Just before an important race this blood could be given back to the athlete—a transfusion. He would then have more blood to carry more oxygen.

Marathon running, over the strange distance of 49.195km, does seem to need superhuman abilities. Approved refreshments (mostly sugar solution) are supplied after 11km, then at 5km intervals. This means that used glucose can be replaced. The athlete's main problems may, however, be dehydration (drying up) and cramp.

Lasse Viren

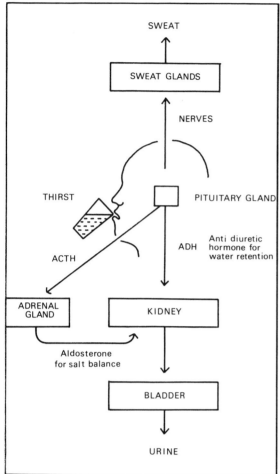

Under normal conditions, our water content is controlled:

water IN = water OUT
food + drink   urine + sweat

The diagram shows how the body controls salt and water balance by means of hormones. This is a good example (simplified!) of self-regulation.

In hot conditions, we sweat more and produce less urine. However, unlike panting dogs, we lose salt as well as body heat and water when we sweat. (You can show this by simply tasting your own sweat, or by evaporating some on a microscope slide, or by mixing some with silver nitrate solution.) This salt has to be replaced so that the body fluids stay the same. Otherwise, cramp may result, or actual collapse. Drinking pure water is all right in small amounts, but large quantities should be slightly salted.

water + salt = normal
no water = fainting, collapse
no salt = cramp, collapse

The watering stations in between the feeding points in the marathon may cool the athletes, but they may not be enough to maintain normal body function. Marathon runners reach towards the limits of human performance.

Long Jump

The long jump can be divided into four stages:

(a) approach, (b) take-off,
(c) flight, (d) landing.   Identify these stages in the drawing.

Approach

The faster the forward speed 'off the board', the longer the jump. So an athlete should really give himself enough length of run to reach top speed. But this 50m or so is too far for accuracy, so 20 or 30m is more common, to ensure hitting the board.

Take-off

The second important factor is the upward speed at take-off, in a line from the thrusting foot through the centre of gravity.

Flight

Immediately after take-off the body starts to roll forwards; any technique which turns the body against this should give a longer jump. The three flight styles (sail, hang and hitch-kick) are shown. As might be expected, the sail is easiest, but least effective; the hitch-kick the hardest to perfect, but the most efficient in increasing the length of jump.

Landing

Whatever the style, the landing is made in a position with the head and feet thrown forward, the body arched, arms thrown in front. At the actual point of landing, some forward rotation is needed to stop the jumper falling backwards, as it is the rear-most mark which is measured. Ancient Greek long jumpers carried weights to correct the rotation of the body. This is not allowed in modern athletics.

High Jump

There are three factors to consider in high-jumping:

(a) raising the centre of gravity as far as possible (H2);
(b) keeping the centre of gravity as near the bar as possible (H3);
(c) clearing the bar with the trailing limbs.

HEIGHT OF JUMP = H1 + H2 - H3

In the discussion of centre of gravity in Chapter 2 (page 17) we have seen how different techniques affect factor (b). You can see how the centre of gravity ideally moves in a parabola. It also gives some idea of the importance of a high jumper having a high centre of gravity (i.e. being tall and thin), and of clearing the bar by as little as possible. What height of jump would be given to the jumper who raised his centre of gravity by only 0.74m?

Most important in the actual jump is the vertical speed at take-off. This depends on the force exerted on the ground, and the time for which the force is applied. As with the impact force on a ball (Chapter 5, page 56), there is a direct relationship; greatest speed is given by the largest force acting for the longest time.

But, even before the athlete leaves the ground, he is already developing the 'lay-out' style for clearing the bar. The upward motion is converted into a suitable body position, before the high point is reached, and the downward path begins.

Throwing

The four throwing events—discus, hammer, javelin and shot—have certain features in common. The important factors are similar to those already considered in relation to the flight of balls. (See Chapters 5 and 6 for more about turbulence, drag, streamlining and lift effects of spin. See Chapter 10 for more about aerofoils.)

Factors affecting throwing events:

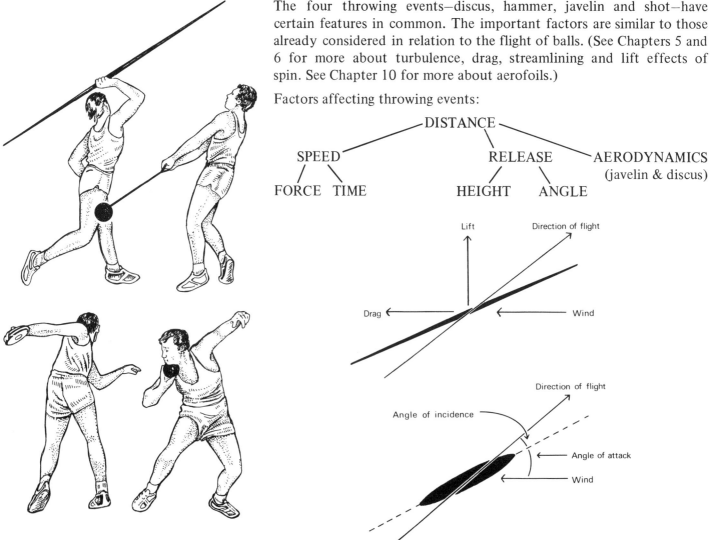

In the best throws, the largest possible force is exerted for the longest possible time, in the direction of the throw. All the various aspects of throwing technique are designed with this in mind. The outcome is a high speed of release, at the most effective angle.

As with balls, the ideal angle of projection should, in theory at least, be 45°. But, while this might be so for projection from the ground, it does not apply to objects which are thrown from 2m or so above the ground. The angle of throw for maximum distance is reduced by this factor. Its actual value depends on the exact height of throw, and the speed of release, but it is about 40° for shot putt, about 30° for javelin.

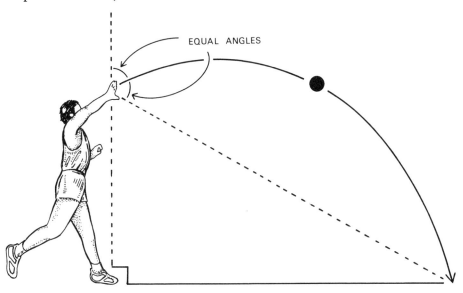

The amount of lift on a flying discus or javelin depends on:

(a) its shape, which affects air flow;
(b) its angle of attack: the larger the better;
(c) its surface area: the larger the better;
(d) its speed: the faster the better;
(e) the air itself—dense air gives more lift.

Of all these, the speed of release appears to be the most important.

Mexico '68: Athletics at 2200m

Two features remain in the memories of those who saw the 1968 Olympic Games in Mexico:

(a) The effects of the high altitude (2200m above sea level);
(b) Bob Beamon's incredible long jump.

Even Britain's two medals in the 400m hurdles, with David Hemerey breaking the world record, were overshadowed.

To long jump 8.90m when the world record then stood below 8.50m was truly remarkable. Beamon himself never again jumped anything near this distance, and no-one else has even approached it. Many experts expect it to survive into the next century. Olympic records were also broken in most of the sprint events, such as the 400m hurdles already mentioned.

Bob Beamon

But there was never any likelihood of records in the distance events. The high altitude of Mexico City means that the air is less dense than that 2,000m below at sea level. In fact, the air in Mexico City is about 77% density, compared with sea level. This makes jumping and sprinting easier, as there is less air resistance to overcome. And, as respiration in these events is mostly anaerobic, the reduced oxygen level is not important. (Gravity is only about 0.05% less, so this difference is not important.)

Because of the reduced air density, there is much less oxygen available. More anaerobic respiration than normal takes place, even in the long distance races. This results in the production of more poisonous lactic acid. High lactic acid content of the blood results in fatigue and collapse. Many athletes had to be given extra oxygen at the end of their races to overcome the unusually large oxygen debt.

Only those athletes who had spent long periods at high altitudes were able to compete at near normal standard. Living in such conditions leads to an increase in the numbers of oxygen-carrying red blood cells, and hence more efficient oxygen transport. It was no coincidence that all the distance races were won by native altitude dwellers (e.g. Ethiopians) or by athletes who had spent long training periods at altitude. Other athletes suffered from the same sort of altitude sickness common to mountaineers.

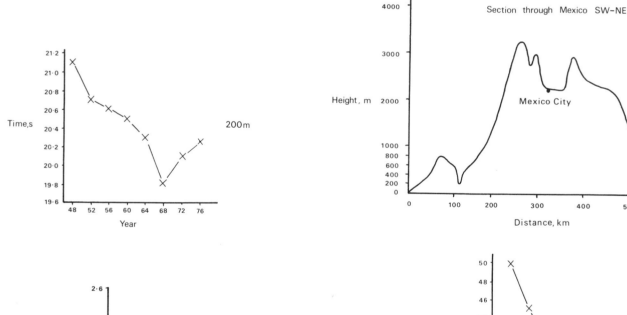

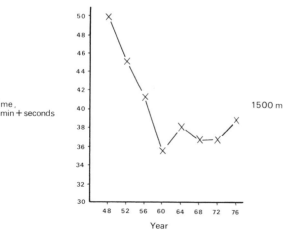

The odd results of Mexico show up clearly if graphs are plotted of Olympic winning performances. Most are similar shapes, a flattening curve, apart from occasional peculiarities caused by exceptional athletes. A change in technique or equipment may also affect the pattern. But most oddities are in the 1968 results.

Questions

1. How does the sprint start increase speed?
2. What is meant by 'oxygen debt'?
3. What percentage of respiration is *aerobic* in (a) 100m, (b) 800m, (c) marathon?
4. State four important features of hurdling.
5. Name four substances vital to distance runners.
6. Which aspects of jumping are shown here?
7. What is the best angle of throw (a) in theory?
   (b) for shot putt?
   (c) for javelin?
8. Copy and complete this paragraph:
   The 1968 Olympic Games were held in .... at an altitude of .... above sea level. At this altitude, air is only ...... as dense as at sea level. This may have helped .... to make his historic long jump of ....m, but it hindered distance runners. Several medals in these races were won by athletes from .... who were used to living at high altitudes. Their blood contains more ....

Extra: Technology in Athletics

'New track will smash world records'

'Carbon fibre answers pole-vaulters' prayers'

These are the kind of headlines which newspapers have used to announce recent advances in sports technology. Many improvements in athletic standards have been due to advances in physiology: diet, training methods, better understanding of endurance and stamina. Some have simply been due to changes in the rules: hurdlers were not originally allowed to knock over any barriers. We are more recently, however, seeing an impact of technological development.

Look again at the changes in the world pole-vault record on page 31.

Poles have, at various times, been made of bamboo, aluminium and glass fibre. The introduction of the glass fibre pole had a marked effect on the heights jumped.

Glass fibre is an interesting material. It is known as a 'composite', made up of two different materials. Just as bones are made of a group of soft, flexible fibres in a matrix (glue) of inorganic material, so glass fibre is made of fibres of glass embedded in a plastic matrix. The resulting composite has the strength of one material, and the flexibility of the other.

The latest development uses fibres of carbon, and promises to be even better than its predecessor.

10 Mechanical Sports

In certain sports, the human athlete is of much less importance than a machine. We can think broadly of three kinds of sport. (1) Those which need nothing other than the 'pure' human participants, such as running and wrestling. (2) Those which need a few simple items, such as cricket or tennis, high jump or shot putt, fencing or weightlifting. (3) Those which depend totally on machines. In this chapter we shall look at five of these: motor-racing, sailing, cycling, gliding and rowing.

Motor Racing

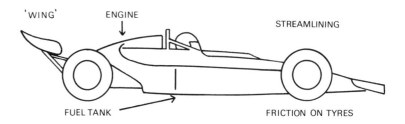

The drawing shows the aspects of a racing car which we can look at fairly simply in scientific terms. The first is the conversion of energy, from chemical energy to kinetic energy, from fuel to power.

The production of movement in an engine is a bit like the process which occurs in our muscles. Carbon and hydrogen bound up in the fuel are burned in oxygen to give carbon dioxide, water and large quantities of energy. This energy appears as useful kinetic energy, and also as waste heat energy.

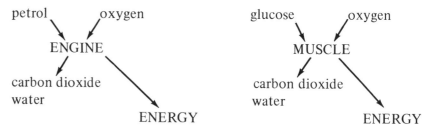

The forces acting on a racing car are complex; some are shown in the diagram:

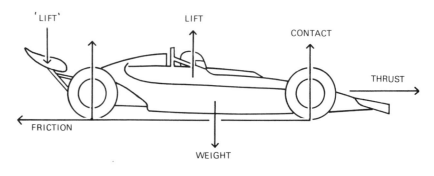

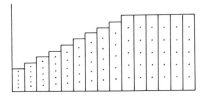

The engine produces thrust, which pushes the car forwards. This is opposed by friction, acting mainly between the tyres and the track, but also within the car itself. There is also air resistance (or 'drag') working against the thrust. The weight of the car (produced by the force of gravity) is opposed by the contact force. The aerofoil design reduces drag, but tends to lift the car. Recent developments have seen, firstly, the 'wing' which gives downward 'lift'(!) and then a fundamental re-design of the car to provide a net downward force (known as 'ground effect'). All the forces must balance in favour of thrust, if the car is to travel forward well and efficiently at high speed.

A car moving at constant speed in a straight line has its forces balanced. If it is well designed and well-made, there should be no difficulty for the driver. As soon as either the speed or direction changes, other forces come into effect. More stress is put on the car. Problems, including accidents, happen when cars are accelerating, cornering or braking. Two investigations will give some idea of these difficulties, and the extra forces involved.

1 Set up the track so that the trolley runs freely along it, without needing a push.
2 Clamp the ticker-timer at the top of the track.
3 Attach the tape to the trolley, making sure that it will move freely through the timer.
4 Switch on, and release the trolley from the top of the track.
5 Look closely at the pattern of dots on the tape. Cut it into convenient portions, every 5 (or 10 or 20 etc.) dots—*not* equal lengths.
6 Stick the pieces side by side in the right order, as in the example shown.

The 'graph' which this produces shows the change of speed caused by gravity. Scientists call the 'the acceleration due to gravity'. Its value is just under $10 m/s^2$. This means that, under the influence of gravity, an object's speed will increase by 10m per second every second. (This is written as $10 m/s^2$ or $10 ms^{-2}$.) Of course, the trolley will not accelerate this quickly because of the friction of the track.

You can get other graphs by pushing the trolley, or by stopping it suddenly, or by colliding it with another trolley, or in many other ways. A straight, horizontal line always shows constant speed (no acceleration); a line sloping downwards (from left to right) indicates deceleration, or slowing down. By changing the surface of the wheels and/or the track, you can investigate the effect of friction on speed and acceleration.

For the second investigation you will need to beg or borrow (do not steal!) a set of model electric racing cars.

1 Find the mass of a car.
2 Use a small forcemeter to find the force needed for the car to just move.
3 With the mass in kg (from g, divide by 1000) and the force in N, the acceleration is given by the equation

$$\text{Acceleration} = \text{Force} \div \text{Mass}$$
$$m/s^2 \qquad N \qquad kg$$

4 Set up a simple circuit, and find the highest speed at which the car stays on the track while cornering. This is best done with photo-

cells and an electronic timer, but can be done with a stopwatch.
5 Try to find ways of increasing this speed: banking the bend, adding mass ('Plasticene') to the rear of the car, roughening the tyres to increase friction, etc.

Cycling

The modern racing cyclist has very little area in contact with the ground. Balance is therefore a major problem, made worse by the small mass of the machine, compared with the rider. This results in a high centre of gravity, and low stability. When the cyclist is cornering, the forces in operation are complex.

By calculating the forces involved it can be shown that the maximum possible cornering speed depends on the radius of the corner and the angle of the lean. This angle is, however, limited by the amount of friction between the wheel and the track. Tight corners or cinder tracks cannot be taken as quickly as gentle bends on tarmac. On banked corners, the slope of the banking becomes the most important factor.

Gliding

Recent racing cars have made much use of aerofoils in their designs. Firstly, they were used simply to reduce drag; later they were designed to keep the high-speed cars on the track. Gliders use the aerofoil principle for the opposite effect: for true, upward lift. It's quite easy to show how the aerofoil produces such lift:

1 Take a sheet of paper.
2 Fold it in half, but do not crease it.
3 Stick one end a few cm. in from the other end.
4 Make a hole through the centre of each half.
5 Pass a thread through the holes, so that it can move freely.
6 Hold the thread vertically between thumb and fingers of each hand.
7 Spin around, or run along, holding the pointed end of the wing in front.
8 Notice what happens to the 'wing'.

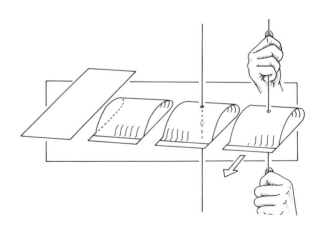

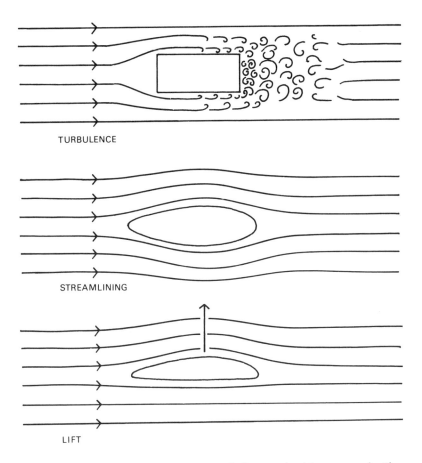

TURBULENCE

STREAMLINING

LIFT

The principle of the aerofoil is straightforward. Air passes both over and under the wing. The air over the curved top has further to travel than the air under the relatively flat base. If the air flow is to be streamlined, and the wing is to stay stable, these two parts of the air must reach the back of the wing at exactly the same time. The air passing above the wing must, therefore, travel faster. Faster air has less pressure. With less pressure above the wing than below it, the wing must rise.

(If the wing is upside down, the air going underneath will have greater speed, and less pressure. The wing will then move downwards. This is what happens in the racing car design.)

Like albatrosses and buzzards, gliders make use of 'thermals' for gaining height. These thermals are currents of warm air; as warm air is less dense than colder air, it tends to rise. A glider may rise hundreds of

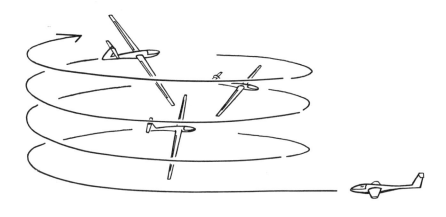

metres on such a thermal, and a skilled pilot can make use of them to stay up for several hours. The fact that thermals are relatively narrow means that the glider has to circle to stay in the warm updraught.

(The longest recorded flight lasted over 15 hours.)

The forces acting on a glider are similar to those on the racing car, although their sizes and their balance are rather different.

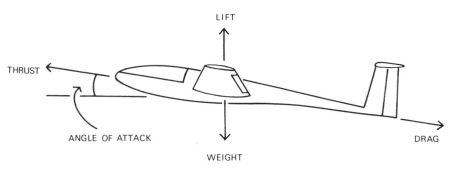

Sailing

It may surprise you to discover that sailing depends on the same basic principles as gliding. The aerodynamics of the movement of air over a curved surface this time apply to the sail. The problem again is similar; to produce enough thrust to overcome the (very large) drag, of the water and the air itself. Solving the problem relies on both hull and sail design.

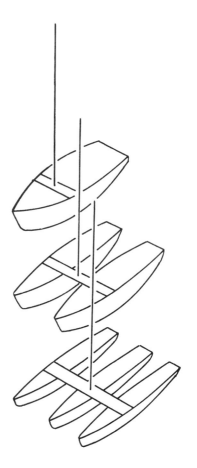

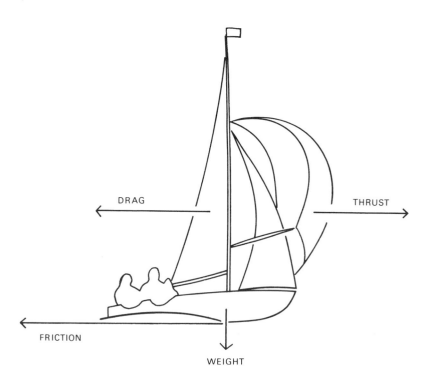

Drag depends on the shape and speed of the boat, and the amount of contact between boat and water. New materials such as fibre-glass and carbon-fibre have improved the shape of boats. However, the increased speed which they produce only increases the drag again. One way of overcoming this is to reduce the contact area: hence the catamaran (double) and trimaran (triple) hulls.

Early sailing boats had a single, rigid, square sail. This meant using the wind at right angles to the sail to push the boat along. Most children draw boats like this. Such boats cannot, however, sail across or against the wind. A manoeuvre such as 'tacking' uses aerofoil principles:

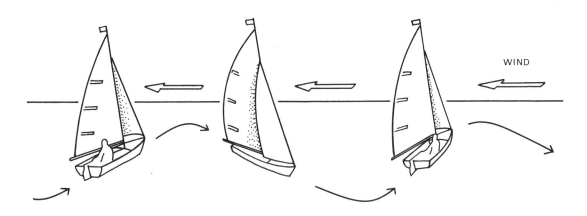

Rowing

Rowing consists of power strokes and recovery strokes. During the power stroke the oars move backwards, relative to the boat. During the recovery stroke the oars move forward.

For the greatest efficiency, the speed of the oars must not be much

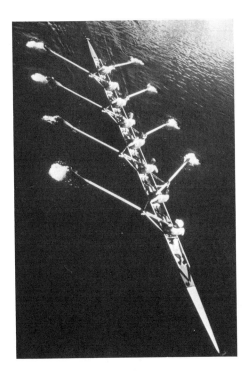

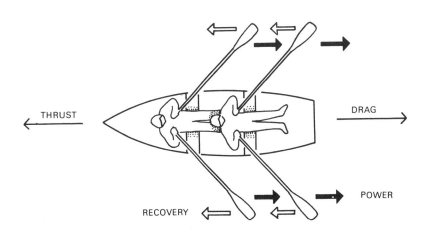

greater than the speed of the boat. And, while the drag on the boat must be as small as possible, the drag on the oars must be large. For this reason, they have large, flat blades. In the recovery stroke, drag is reduced by taking the oars out of the water. Even the drag of air resistance is reduced by turning the blades horizontal during recovery. This is known as feathering.

Questions

1. In what ways is a car engine similar to a muscle?
2. List the forces acting on a racing car.
3. Look at these results from a ticker-timer experiment.

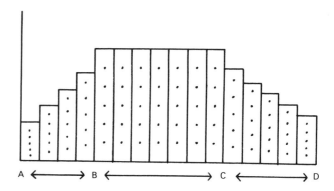

What happened from (a) A to B?
                          (b) B to C?
                          (c) C to D?

4. Explain the principle of the 'wing'.
5. What is a 'thermal'? How is it used by a glider?
6. Name two modern hull materials.
7. What are catamarans and trimarans?
8. Describe the movements of an oar in rowing. How do these movements affect the amount of drag?

Extra: Man-powered Flight

Man-powered flight has long been a dream. Ever since Icarus, men have tried to fly. Most of the early attempts failed because they tried to copy the wrong thing. They tried to build systems which mimicked the flapping motions of the wings of birds and insects. Really, flight is concerned with forward thrust, and an aerofoil shape to convert some of this into lift.

The essential problems of flight are

(a) a design to give as much lift as possible;
(b) producing enough power to lift the weight of the body plus the weight of the craft;
(c) being able to maintain the necessary power output for sufficient time, so that 'flight' is achieved, rather than an extended jump.

The overriding difficulty is that increasing the power potential of a man-powered system means a more muscular athlete, and this in turn means more body weight to be carried.

The two graphs give an idea of the problem. Notice the scales on the graphs: three of them are logarithmic, with each mark on the axis being 10 times the value of the previous one. More details about such graphs can be found in the Extra section on page 38.

From the graphs you will see that Man is just on the borderline of the power/weight ratio needed for steady flight, and that even a champion athlete would have trouble in maintaining this rate of work for more than a few seconds.

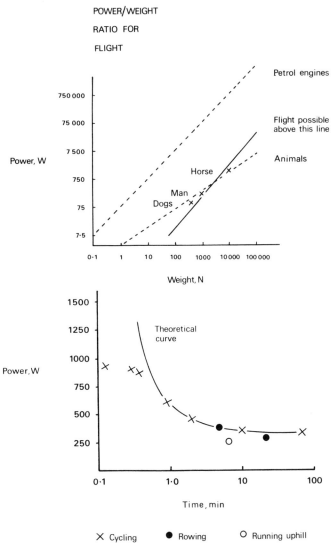

But this is theory. Some twenty years ago, a prize worth today about £60,000 was offered to the first man to make a man-powered flight around a prescribed figure-of-eight course. In 1977, the 'Gossamer Condor' finally succeeded, on August 23, in making designer Paul McCready richer by that amount. Cyclist Bryan Allen carried almost 93kg (912 N) of which two-thirds was his own body. The later 'Gossamer Albatross' is shown crossing the Channel.

Gossamer Albatross

11 Playing Surfaces

Ball Bounce

We have seen (Chapter 5) how the nature of the ball can affect bounce, roll and spin. The playing surface and weather conditions are also important. Two of the most popular summer games—cricket and tennis—supply good examples.

Lawn tennis, as its name suggests, was at first played only on grass. Today, there is a variety of surfaces: grass, concrete, clay, cinders (all outdoors); wood and synthetic plastic materials (mostly indoors). Each of these has its own features:

Surface	Bounce	Effect of rain
Grass	fast, low	slippery, ball skids
Concrete	fast, high	drains
Clay	slow, high	dries quickly
P.V.C.	slow, high	dangerous
Nylon	fast, low	dangerous

Table 11.1

Cricket talk is full of phrases about the state of the pitch: 'sticky wicket', 'taking spin', 'green on top', 'caught on a drying wicket'. There is, in fact, a surprising amount of 'give' in even the hardest pitch. This elasticity means that the ball bounces higher than might be expected, and slows down. On a soft wicket, the ball bounces awkwardly as it rises and slows.

If you repeat the bounce experiment, using the same ball on different surfaces, you will get some idea of the effect that the surface has. Sample results are shown in the graph.

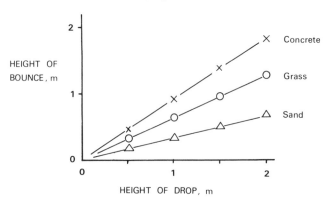

Running Tracks

Running tracks, like tennis courts, may be made from a variety of materials. Each of them has its advantages and its disadvantages.

Grass

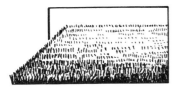

is common in schools. It has advantages in that it allows different kinds of footwear, is fairly easy to maintain, and can be taken over by other pitches when not used for running. Its major problems are that it is very slippery when wet, and may get worn away to bare soil.

Tarmac

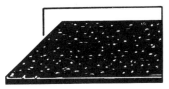

drains and dries easily. Athletes are, however, only able to wear soft shoes, and can be badly hurt in a fall.

Cinders

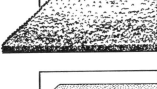

is the traditional surface. It combines durability of tarmac with a softer surface, but is easily waterlogged.

Synthetic materials

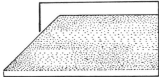

such as the 'tartan' track, have recently been developed in an attempt to combine the best features of all the other types. There have, however, been problems with these.

In the Montreal Olympics, there were several accidents in the distance events. These were blamed on the fact that the track was 'silent'; runners could not hear others close behind them. And a tragedy was narrowly avoided when a hammer bounced off the track near a group of runners.

There is more about new tracks in the Extra section.

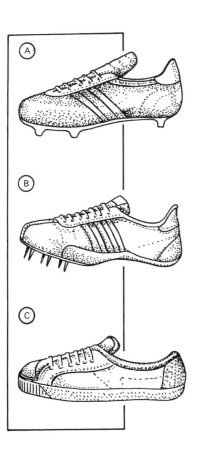

Footwear and Friction

Sports footwear must provide (a) support
                              and (b) suitable friction.

There are three basic types of sports footwear:

(A) studded boots—soccer, rugby etc.;
(B) spikes—athletics;
(C) tennis/training shoes—hard surface and indoor games.

Abnormal weather conditions (such as winter 1979) and artificial surfaces (such as Astroturf) have served to highlight the importance of suitable friction. Too much friction makes movement difficult. Too little friction and the slightest change of speed or direction becomes dangerous.

Friction between footwear and surfaces can be estimated by dragging the shoe or boot across the surface with a forcemeter. The frictional force is read just as the first movement occurs.

As with the cornering cyclist, the turning sportsman on foot can only lean so far. The angle of lean depends on the amount of friction between his contact foot and the ground.

108 SPORT SCIENCE

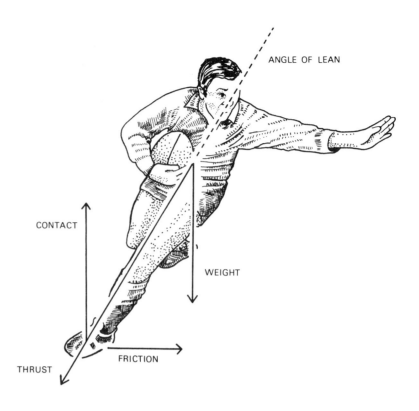

Wet weather tyres

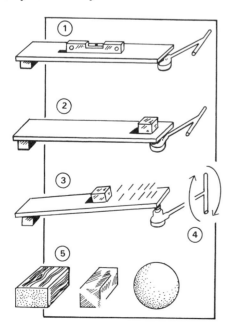
Dry weather tyre

Friction and Tyres

Friction is a doubly important force in motor racing. In the engine, and other moving parts, it is an enemy, fought with lubricating oils. On the track, as the wheels spin, just the right amount is needed for the tyres to grip the track. Too little, and the car slips and slides as if on ice. Too much, and the car slows down.

Racing cars normally have several sets of tyres, for use in different weather conditions, and on different surfaces. If the track is wet, tyres with deeper treads are used. If it dries out during a race, there is a rush to the pits to change to smoother tyres.

Winter Sports

Nowhere is friction more critical than in winter sports.

The most important factors in winter sports are friction, balance, and the ice or snow itself.

Ice has two useful properties:

(a) low friction;
(b) it melts under pressure.

Both of these can be shown quite simply by two experiments.

1. Support a plank on a jack and block so that it is perfectly level.
2. Put an ice cube on the plank at the jack end.
3. Carefully raise the jack until the ice slides.
4. Measure the number of turns of the jack, and/or the angle of the plank.

5 Compare the results with those of other materials, such as wood and metal—and with balls (Chapter 5).

1 Clamp a block of ice firmly.
2 Attach weights to a length of wire.
3 Place the wire across the block.
4 Watch carefully.

The pressure of the loaded wire melts the ice, and it moves downwards through the ice; above the wire, the pressure is less, and the ice re-forms.

This effect is made use of in both skating and ski-ing. The very thin edge of the skate blade produces a large pressure. (Pressure increases as force on a given area gets larger or area gets smaller.) Heavy skaters and thin blades both produce larger forces.

$$\text{Pressure} = \frac{\text{force}}{\text{area}} = \frac{\text{weight of skater}}{\text{blade length} \times \text{width}}$$

length of blade (cm)
×
thickness of blade edge (mm)
÷
100 000
=
M
C
weight of skater (N)
÷
MR
=

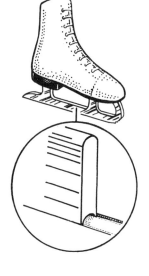

This pressure compares with about 50 000 N/m² for an average man, 200 000 N/m² for an elephant.

You should find that, even for a light (600 N) skater on the full blade (3mm × 30cm), the pressure is more than the elephant. If the skater is skating a figure, the edge of the blade may be as thin as 0.1mm. This gives an enormous pressure.

The very high pressure melts the ice, and means that skating takes place on a thin film of water. This has extremely low friction. There must, however, be some friction; otherwise, there is no thrust, and no movement.

Thrust in ski-ing is provided by the sticks/poles. Friction is reduced by rubbing wax underneath the skis.

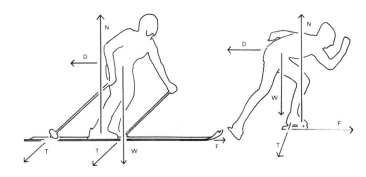

Wear and Tear

All surfaces wear away. Some, like grass soccer pitches, wear a lot. Others wear only a little. There are two main factors:

1 the material of the playing surface;
2 the spread of activity.

Two examples will illustrate this:

On a running track, the lanes are used unevenly:

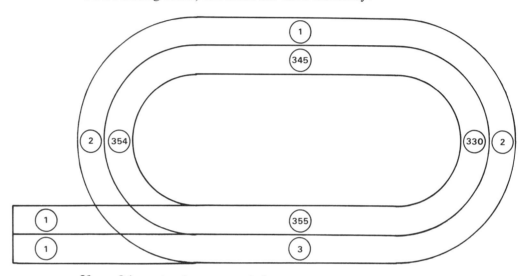

If an 8-lane track was used for one race at each distance, each with 8 runners, the equivalent of 355 athletes would run on the inside lane of the 'home' straight. The outside lanes of the 'back' straight would be used only once!

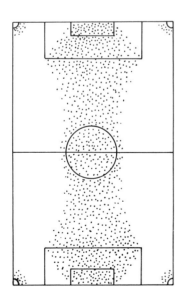

WEAR ON A SOCCER PITCH

Race, m.	Lanes used
100	All, 'home' straight only
200	All, 2nd half of circuit only
400	All
800	All for first bend, then inside lane only
1500	Inside lane only (except overtaking)
5000	Inside lane only (except overtaking)
10000	Inside lane only (except overtaking)

By the spring, most soccer pitches look rather the worse for wear. Analysis of the main areas of activity (Chapter 4) would confirm the evidence of wear. Most play occurs in the penalty areas and centre of the pitch. The diagram shows the most heavily worn areas.

Questions

1. What happens to ice when pressure is applied to it?
2. How much pressure is produced by a 500N ice-skater on a 25cm x 0.1mm blade edge?
3. State three facts about friction in winter sports.
4. List the factors which decide whether a swerving rugby player falls over.
5. Why is the amount of friction between foot and ground important?
6. What is 'Astroturf'?
7. (a) What kind of bounce is produced on a grass tennis court?
   (b) What effect does rain have on this bounce?
8. What effect on the bounce of the ball is produced by the elasticity of a cricket pitch?

Artificial Ski Slope

Extra: Playing Surfaces and the Weather

The winter of 1979 did more than make soccer players learn to skate, or to change their boots during matches. It raised the whole question of weatherproof and/or artificial pitches. Leicester City's 'hot air baloon' system not only protected their own matches, but also enabled the Filbert Street ground to stage three of the Arsenal v Sheffield Wednesday F.A. Cup replays. Somewhat ironically Arsenal's own under-soil heating was at the time one of the few other protective measures taken by any club. Such a system also keeps grass on the pitch all the year round.

Astroturf (named after the Houston, Texas, Astrodome) is a nylon material. Although approved by F.I.F.A., and easily swept clear of snow or rain, many players in Britain have not been keen on it. Several clubs have, however, now installed practice areas of such material. With top-class tennis now played on artificial surfaces outdoors as well as indoors, perhaps grass pitches will soon be found only in history books.

Where the weather is too warm for snow ski-ing, artificial ski slopes have been built. These are usually made of tufted nylon (like a brush), and provide suitable friction for learning to ski, or even for advanced skiers to practise on. Most recently, a 'moving carpet' (ever upwards) has removed the need to repeatedly climb to the top!

A New Running Track

Running tracks have changed, too. We have seen the materials used in the past. The most recent design (at Harvard, U.S.A.) is designed to 'bounce' in harmony with the runner: A rubber covered wooden suspension system supports the fibre-glass and steel track. Because it 'gives' with the runner, the contact time is greater. A greater impulse force results, and therefore more speed. A second advantage is that the shock of landing is reduced, and fewer injuries occur, particularly knee injuries. It is estimated that the Harvard track could reduce middle-distance records by several seconds.

12 Practice, Training, Injuries

Practice makes perfect?

There is often argument about whether sportsmen are born or made. Some people certainly have better 'ball sense' or excellent muscle co-ordination. Some are physically stronger than others; some can run faster. The question we need to answer is, 'How far can practice improve performance?'

Practice has two separate aspects: results and techniques. We can practise a sports skill so that our results improve, despite a strange technique. Or we can try to perfect our technique, so that better results should follow eventually, if not at once. Here are some simple ways of investigating the effects of practice.

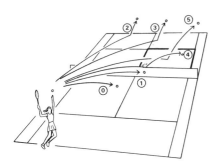

Tennis Serve
1. Serve a number of balls on a tennis court.
2. Score points for each serve that:

bounces before the net	0
lands in the net	1
is over, but long *and* wide	2
is over, but long *or* wide	3
is a 'let' service	4
is a good service.	5

3. Draw a graph of the scores for each attempt.
4. Try again, on other days. Find out whether there is any lasting improvement.

Hockey Dribble
1. Place a series of posts or flags in a straight line at, say, 3m intervals.
2. Record the time it takes to dribble, slalom-style, along the 'course'.
3. Repeat several times.
4. Draw a graph showing the time for each attempt.

Darts
1. Set up a dart board.
2. Count the number of throws needed to throw a '1'.
3. Count the number needed for a '2'.
4. Continue, in turn, from 3 to 20.
5. Draw a graph to show the number of attempts at each number.

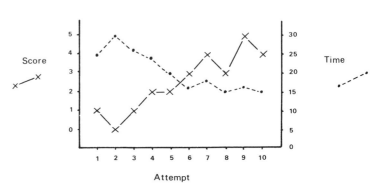

These are just three examples of the many kinds of practice measurements that can be made. Your graphs may look like the one on page 112.

Unless the test subjects are already 'experts' at the practice task, there will be an improvement in the early stages. Then the graph levels out, as peak performance is reached. If tiredness becomes a factor, the results may even begin to get worse.

The examples are all based on the first aspect of practice: improving results, without considering technique. You could try some coaching of technique between attempts, to see whether this affects the results. If we are worried about technique, rather than results, a different approach may be helpful.

Experimentally, the difficulty is to choose a skill the subjects don't already have. One way is to try to develop a technique using the 'wrong' hand (or foot). Right-handed people should use their left hands, and vice-versa. Practice may now need to be over a much longer period before effects are noticed. And many other factors may be found to influence the results:

> noise
> tiredness
> possession of similar skills ('transfer')
> competition
> understanding the mechanics of the task
> not knowing the result of each attempt

Complex skills like swimming take a long time to master, but practice and coaching of actual techniques certainly help. Here, analysis of movement (Chapter 4) may give useful information. Many athletes will not appreciate how odd their technique is unless they can see it for themselves.

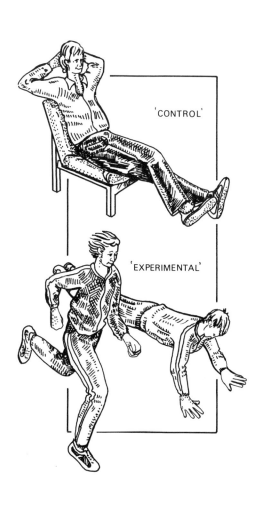

Training

The terms 'practice' and 'training' are often used as if they mean exactly the same thing. Certainly, the techniques of training—squat thrusts, weights, chins etc.—can be practised, like any technique. But the emphasis in practice is on skill and results. Training concentrates on physical fitness.

Measuring the effects of training is difficult. We need to compare people who train with people who do not. They must, however, be equally fit to start with; otherwise our results will be meaningless. We have to be certain that any changes are due to the training. Do not attempt this experiment without qualified supervision.

1. Collect a group of 'volunteers'.
2. Record the height, mass and pulse rate of each volunteer subject.
3. Measure each subject's index of fitness, by means of the Harvard Step Test (page 82).
4. Divide the group into two. Try to match the two halves for age, sex, build, fitness etc.
5. Record some other fitness measure as a check. Standing high or long jump, time to run 1 km, number of press-ups or similar will do.

6   One group does no training: this is the 'control' group.
    The others should do some kind of physical activity (running, press-ups etc.) every other day for a month: this is the 'experimental' group. Discuss the details with your supervisor.
7   After a month, repeat all the measurements with both experimental and control groups.
8   Draw up tables and graphs of the results to see whether the training has had any effect.

Other training/measurement exercises which you can try include:

   shuttle-runs
   pull-ups (chins)
   50m sprint
   ball throwing
   weightlifting (under supervision only)
   squat thrusts
   swimming

Physiological Effects

The first training session at the start of a season usually leaves aching muscles. It may even be difficult to walk, especially down stairs. Many muscles will have done little work for a long time. As training becomes more frequent and regular, these aches vanish, and the muscles become more efficient.

Not so obvious as ordinary limb muscles are those of the heart. Heart muscle increases in strength and efficiency as a result of regular exercise. The volume of the heart also increases. This means that the same amount of blood can be supplied with fewer beats. The heart of a trained athlete therefore beats less often, but more strongly.

cardiac output   =   stroke volume   ×   heart rate
(blood pumped   (blood pumped   (beats per
per minute)   per beat)   minute)

You can check the truth of this by means of a simple survey.

1   Take the resting pulse rate of a group of people, the same age and sex.
2   Record each in a table according to whether they
    (a) never take exercise,
    (b) take occasional exercise,
    (c) take regular exercise.
3   Find the average pulse rate for each group.

Sample results:

Group	a	b	c
Pulse rates (beats/min)	84 80 76 80 80	60 72 76 66 62	44 52 60 46 48
Average			

Table 12.1

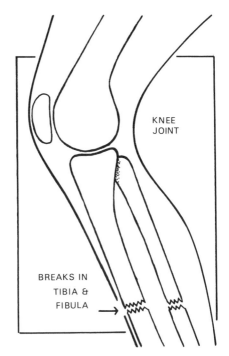

And, of course, the pulse rate of a fit person returns to normal much sooner after exercise, as we have made use of in the index of fitness calculations.

Injuries

Injuries are an accepted hazard in most sports. In some (motor-cycling) they may be fatal. In others (boxing) they are inevitable. In most sports they are simply more or less frequent, more or less severe.

Most injuries are a minor, temporary inconvenience. A few cause the sportsmen to retire. Those in between may be the most interesting.

One of these is the double fracture of the leg. ('Double' here means both tibia and fibula—not both legs!) It is most common in soccer. This injury may take a year to heal properly, and the player is 'never quite the same again'. An unusual case was that of Ian Evans (Crystal Palace and Wales). One bone healed, the other did not. A new treatment at the London Hospital was tried. This involved applying a constant electrical potential across the break for three months. Success finally permitted a return to action after 2 years on the sidelines.

Cartilage

is the material which protects the ends of bones from damaging each other. But if it is torn, so that it becomes broken or separated, it may need to be removed. This most common knee operation is usually spoken of simply as a 'cartilage operation'. Damage happens most frequently in the knee as it is the knee joint which is most strained by running, jumping or skating. Olympic champion Robin Cousins is one of many sportsmen who have had this operation.

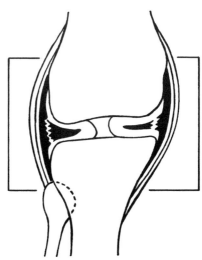

CARTILAGE DAMAGE IN KNEE

Cartilage damage may cause the knee to lock or to give way, and it may click when moved. Swelling and pain are also symptoms.

Fractures

are the most dramatic kind of injury. They can be classified in two main ways:

    A closed : skin stays intact
    B open   : skin broken

    a transverse : break straight across
    b oblique   : break at an angle
    c greenstick : break only part way across

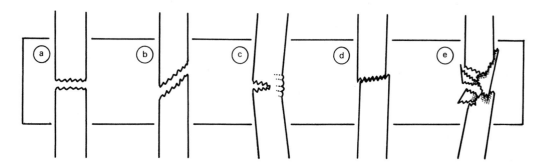

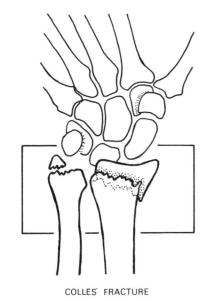

COLLES' FRACTURE

d impacted : pieces locked into each other
e comminuted : more than two pieces.

Fractures may be caused by

(i) a direct blow, e.g. a kick
(ii) an indirect blow, e.g. falling and landing on the heels can fracture the spine;
(iii) muscular contraction, e.g. a split patella while jumping.

Most common fractures during sport are:

(a) Colles' fracture of the radius near the hand, caused by landing on the wrist, e.g. goalkeeper;
(b) fractures of the clavicle (collar bone) during a rugby tackle or a fall;
(c) fractures of the tibia and fibula, e.g. a soccer tackle;
(d) Potts' fractures of the ankle, caused by twisting the foot, e.g. ski-ing.

Dislocations

can, like fractures, be dramatic. Here, the positions of the bones in a joint are altered by a blow. Most common dislocations are of the shoulder (rugby tackle or fall) and finger joints (oblique blows by balls).

Sprains and Tears

are common in most sports. A sprain occurs when a ligament is stretched or torn, but not with enough force to dislocate the joint. Sprains are most usual in the ankle, where landing unevenly causes the joint to 'turn over'. Firm bandaging reduces swelling as well as providing support for the sprained ankle. Sprains can be as painful, swollen and bruised as a break. Gentle exercise should be increased as the pain eases.

Muscle Injury

is not always properly named. There are two main types of injury connected with muscles: haematoma (burst blood vessels and clots within the muscle) and so-called 'pulled' muscles.

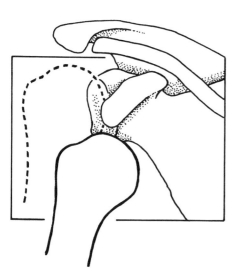

DISLOCATED SHOULDER JOINT

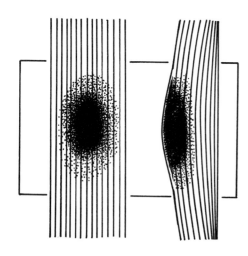

HAEMATOMA

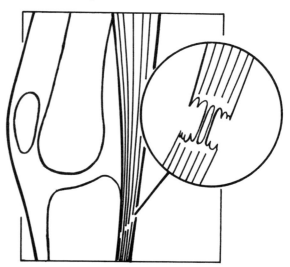

HAMSTRING TENDON INJURY

Haematoma is produced by a sharp blow to the muscle, causing a blood vessel to burst. Blood collects, and a clot forms inside the muscle, making contraction painful. Pulled muscles do occur, but the more usual site of damage is the tendon which attaches the muscle to a bone. This may become stretched, torn or completely separated. A common location for this is the hamstring, at the back of the thigh.

Other symptoms of tendon damage include swelling, thickening or actual splitting. The Achilles tendon above the heel is the best-known for this kind of injury. Others include 'Housemaid's knee' (which may affect goalkeepers) and 'tennis elbow'.

Sports Medicine

One kind of injury not mentioned is that caused by deliberate violence. These are not really sports injuries, and may therefore be of any kind. For example, occasional cases of 'studding' and biting occur in rugby.

J.P.R. Williams

A well-publicised case was that of the Welsh captain and full-back J.P.R. Williams in the 1978-79 season. His badly scarred face was said to be due to a malicious act by a New Zealand forward during a match between Bridgend and the 1978 All Blacks.

The injury to 'JPR' was ironic in three ways. Firstly, he is a doctor, better suited than most to understand the problems of injuries. Secondly, he 'retired' from international rugby that same season. Thirdly, his main field of interest lies in establishing and running clinics for sports medicine.

Such clinics for sports medicine have increased in number in Britain over the last few years. The bar charts show some data from such a clinic at St. James' Hospital, Leeds.

Outside the National Health Service there has been an increase in interest and activity in the field. All Football League clubs, for example, have a physiotherapist on the staff. Some, such as Norwich City, have run courses for local coaches and trainers, in order to spread their specialised knowledge.

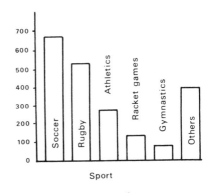

Sports medicine has, of course, more to concern itself with than just injuries. The whole realm of fitness, diet, strength, stamina and so on lies in this field. Britain has had in the past a curious attitude towards the highly specialised work done in, for example, East Germany. There every detail of an athlete's body is examined and recorded, so that training can be adapted to his own particular needs. There is even a system of training-down, so that retired athletes do not gain weight in undesired ways.

First Aid

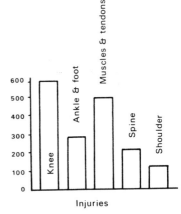

With any injury, rapid action may prevent further damage. On the other hand, the wrong action may soon make matters worse. Often, a sportsman may suffer worse injury simply by playing on; or by making a 'comeback' when not fully fit. It is useful to know not only what to do, but also what *not* to do, when an injury happens. Table 12.2 covers the most common injuries.

INJURY	WHAT TO DO	WHAT *NOT* TO DO
Bleeding	Cover wound with sterile dressing, or best material available. Rest; raise wound. If serious, stop bleeding with pressure to wound, and obtain medical aid.	Do *not* apply a tourniquet.
Nose bleed	Sit with head between knees. Pinch bridge of nose; apply ice or cold compress.	Do *not* tilt head backwards.
Fractures:		
tibia	Splint both sides with anything suitable—corner flags and cricket stumps will do! (Or use the other leg.)	Do *not* move unless splint applied. If in any doubt, cover and call medical aid.
radius	Splint from elbow to knuckles. Make sling from triangular bandage.	
clavicle	Pad under armpit, bandage upper arm to chest, sling forearm towards opposite shoulder.	
Dislocations	Support with bandages or sling.	Do *not* try to replace joint without medical help.
Sprains	Apply cold compress, then firm bandage. Raise joint and rest.	For ankle sprain, do *not* remove footwear, or swelling may make replacement impossible.
Concussion	Rest and medical aid essential.	Do *not* allow to continue or to walk home.
Unconsciousness	Turn onto side. Clear mouth of obstruction. Obtain medical aid as soon as possible. On recovery, force to rest.	Do *not* allow to lie on back. Do *not* give anything to drink. Do *not* allow return to activity. Ignore pleas of 'I'm all right; let me back on the field.'
Breathing stops (& heart)	Artificial respiration (and cardiac massage). Loosen clothing; clear throat of any obstruction. Obtain medical aid as soon as possible.	Do *not* wait for medical aid, get the nearest teacher or qualified lifesaver.
Exhaustion	Cover to keep warm. Force to rest. Give sugar or other simple energy source.	Do *not* allow to continue, however keen.
Exposure	Wrap up warmly. Give sugar or other energy source. Gently massage to encourage blood flow.	Do *not* heat up too quickly. Do *not* overwrap head and neck.
Cramp	Stop activity. Massage.	Do *not* continue exercise.

Table 12.2 Note: Shock may occur following any injury. There is no specialist treatment for shock. Loosen clothing and force to rest. Keep warm.

Infections

Certain infections or diseases are closely associated with sports. Three of them are described here; you may be able to think of others.

Athlete's Foot

will be most people's first thought of a sports-linked disease. Athlete's foot is an infection of the skin between the toes. It is caused by a fungus, a kind of ringworm. It is easily transmitted to others by means of towels, socks, wet changing-room floors. For this reason, swimming pools usually have a disinfectant foot bath outside the changing room, to prevent its spread around the pool.

Verrucas

are spread in a similar manner. In this case, though, the hard lumps on the foot are caused by a virus. This is not much affected by disinfectant, so swimming is not usually allowed while the infection persists.

Impetigo

is caused by bacteria. It affects the face, where unpleasant yellow vesicles form. It is easily spread by facial contact. Occasionally a whole term's rugby has been lost in schools and colleges when an outbreak of impetigo has occurred among prop forwards!

Warming-up

Warming up has been mentioned as a way of avoiding pulled muscles, particularly the hamstrings. It has the function of gently loosening muscles, rather than suddenly expecting maximum effort from them. It also encourages circulation of the blood to the muscles, rather than, for example, the intestines. And, as its name suggests, it does warm up the body. Muscles do actually work better at 38 or 39°C.

Questions

1. These results were obtained during a study of the effects of practice on taking penalty kicks.

Attempt	1	2	3	4	5	6	7	8	9	10	11	12	13	14	15
Result	G	M	G	M	G	G	G	S	G	G	G	S	S	G	G

G = goal;   M = miss;   S = save

Is there any evidence of a practice effect on (a) the kicker, (b) the goalkeeper?

2. What are the two aspects of practice?
3. List five exercises suitable for use in a study on the effects of

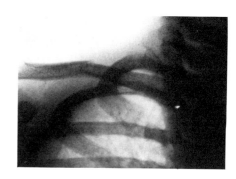

training.
4. What is the effect of training on the heart?
5. From Table 12.1 on page 114, what are the average pulse rates for groups a, b and c?
   What do the results suggest?
6. What injuries were suffered by (a) Robin Cousins, (b) Ian Evans? What treatment did they have?
7. What kind of fracture is shown?
8. Where would Colles' and Potts' fractures occur? How might they happen?
9. What are (a) sprains, (b) dislocations, (c) pulled muscles, (d) haematoma?
10. Who is 'JPR'?
    Why is he most suitable to include in the section on injuries?
11. What should be done for (a) a fractured clavicle, (b) a nose bleed, (c) cramp?
12. State three functions of warming up.

Extra: Smoking and Fitness

Cigarettes and smoking have a peculiar relationship with sport. A casual glance at the lists of sports sponsors might suggest that smoking is an ideal activity for sportsmen. At various times there have been:

Cricket	:	John Player League; Benson & Hedges Cup
Rugby Union	:	John Player Cup
Golf	:	Picadilly World Matchplay Championship
Motor Racing	:	John Player Special (a car)
		Marlborough McClaren (also a car)
Snooker	:	Embassy World Championships

And how many soccer managers can be seen puffing instructions to their teams?

But, in fact, we know that smoking 'takes your breath away'. Few serious athletes smoke during training. There is good scientific evidence that the ability to undertake strenuous exercise is reduced by smoking, however young the athlete may be.

Indirect evidence can be got by simply looking at the products of a cigarette:

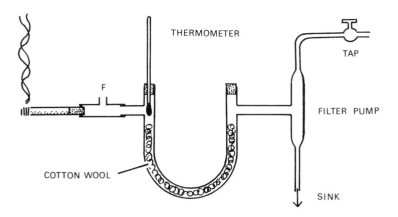

1. Set up the apparatus as shown.
2. When hole F is covered with a finger, the cigarette is 'inhaled'. Cover the hole to light it.
3. Cover and uncover the hole, so that the cigarette is 'puffed'.
4. Look at the cotton wool and the thermometer. What do you notice?

The main ingredients of cigarette smoke are:

(a) Tar. This is the dark brown material which collects on the cotton wool.
(b) Irritants. These cause coughing, and affect the natural cleaning processes of the bronchial tubes.
(c) Nicotine. This affects the nervous system. It is a drug, a stimulant at low doses, a sedative at high doses.
(d) Carbon monoxide has the quickest effects. Normally, haemoglobin in the blood carries oxygen:

$$\text{haemoglobin} + \text{oxygen} \rightleftharpoons \text{oxyhaemoglobin}$$

However, haemoglobin seems to 'prefer' carbon monoxide. Unlike its reaction with oxygen, carboxyhaemoglobin does not split up again. Once formed, the haemoglobin is unable to carry oxygen.

$$\text{haemoglobin} + \text{carbon monoxide} \rightarrow \text{carboxyhaemoglobin}$$

Up to 10% of the oxygen-carrying capacity may be lost in this way in the blood of smokers.

Direct evidence comes from studies on students and servicemen. Examples of results from such studies:

Time taken to complete cycling exercise is reduced by stopping smoking.
Endurance and capacity for exercise is reduced in proportion to the number of cigarettes smoked.
Improvement produced by training is less in smokers than in non-smokers.

Can't even play blow-football now!

13 Sport and Social Science

Science and Social Science

The social sciences—psychology and sociology—are an attempt to apply scientific methods to aspects of human life. They have much to say about sport.

Sociology is concerned with how people behave in groups. Psychology tries to explain the behaviour of both individuals and groups. To do this, both use the methods of observation and questioning. Psychologists can also use experiments.

Measurements in social science are largely subjective, because of the difficulty of knowing exactly what is being measured.

Sport for All?

'Are you a keen sportsman'?

'Are you keen on sport'?

Two almost identical questions, but they have very different meanings. The first refers to taking part, the second could refer only to watching. A person who is keen on squash is probably a player. A soccer 'enthusiast', however, may never have kicked a ball in his life. His keenness may be confined to watching, at his local ground, or on his television set. About ½ million people watch Football League games every Saturday. Many millions watch the F.A. Cup Final 'live' on television.

Sport	Event	Crowd
Soccer	F.A. Cup Final	100,000
Soccer	Brazil v Uraguay (1950)	199,854
Athletics	Boston Marathon	1.5 million
Cycling	Tour de France	10 million

Sport	Country	Number of participants
Volleyball	World	90 million
All	USSR	50 million
All	East Germany	5 million
Angling	U.K.	2.75 million
Soccer (registered players)	U.K.	1.66 million

These numbers, however large, are small compared to the numbers

Jimmy Saville

taking part in sport. For every top-class event which draws a crowd, there are thousands where sportsmen and women take part, solely for enjoyment.

Some single events can attract enormous numbers, even marathon running! The 22km run in Milan (Italy) has as many as 50,000 runners.

In Britain, the 1970's saw a vast increase in two sporting areas: jogging and sports or leisure centres. The image of modern man stepping from bed to car to desk to meal came to be associated with overweight, unfit, heart-attack, early death... Encouraged by such as athlete Brendan Foster and disc-jockey Jimmy Saville, 'ordinary' people donned tracksuits and training shoes. Local councils built facilities for squash, badminton and swimming under one roof. The Sports Council's vision of 'Sport for All' took a step forward.

Sport for the Few?

While many watch and many take part, only a few reach the top level in any sport. International sportsmen, Olympic and World Champions are very few. In order to reach this standard, sport becomes increasingly a full-time activity. Two questions are often asked:

> Should a nation devote large sums of money to its top sportsmen or spread its resources as widely as possible?
>
> Is winning so important that any method is justified?

Answering these questions is complex. Dedicated enthusiasts will get up at 5 a.m. if this is the only time to train before work or school. They pay their own expenses. For most British sportsmen, this continues indefinitely. In America, promising athletes are given 'scholarships' to sports-orientated universities. In Eastern Europe, special schools and the armed forces fulfil the same role. Budding gymnasts may be taken away from home at the age of 7.

In some sports, e.g. soccer, there is a long history of professional sportsmen. In others, professionals have become increasingly common in recent years.

Soccer	Payments legalised in 1885-6. 'Amateur' status abolished in 1974.
Rugby League	Reason for break from Rugby Union
Cricket	Gentlemen (amateur) and Players (professional) for many years
Tennis	Have successively become 'open' in the last few years
Show jumping	
Badminton	
Athletics	Olympics strictly amateur, but moves to 'open'
Rugby Union	Last stronghold of the true amateur amateur?

Where professional sportsmen are allowed, there should be little problem—in theory, paying spectators look after the facilities. It is with amateur sports that the questions arise.

Sport as Industry

Sport is now big business. A Wembley Cup Final can take £½ million in ticket sales. Individual sportsmen receive large sums for advertising. Commercial firms sponsor events in return for large amounts of free publicity. Enormous sums of money move between soccer clubs as players are transferred, the losses set against tax. When Trevor Francis became Britain's first £1 million player, he received £50000 for himself.

The influence of money, particularly the rewards for winning, may have had an effect on the standards of behaviour which we expect from sportsmen. Arguing and cheating are often seen as normal.

Cheating

Cheating in sport takes many forms. The use of drugs is discussed below; another classic case occurred in the 1976 Olympic Games. In the modern pentathlon, a Russian competitor was found to have incorrectly wired equipment in the fencing section, so that a 'hit' was shown without actual contact being made. He was disqualified.

It is sometimes suggested that cheating is less common with amateur sportsmen. The 'professional' foul has, however, spread to all levels, together with the attitude that, 'If you can get away with it, it's all right!' Examples of this are not restricted to soccer; arguing with umpires has become more common in tennis and cricket.

SPORT AND SOCIAL SCIENCE

Drugs

The word 'drug' has many levels of meaning, from the addict's heroin to the caffeine in tea and coffee. The problem in sport is to decide what to allow, and then to find ways of detecting any banned drugs.

A well-publicised case occurred in the 1978 World Cup Finals. Routine tests on selected players were made after each match. One such test showed that Scotland's Willie Johnston had been taking a banned drug: 'Reactivan'. He claimed that it was part of a hay-fever treatment, although it contains the stimulant fencamfamin, as well as vitamins B_1, B_6, B_{12} and C.

Other recent cases have seen Olympic medals withdrawn from competitors after the results of similar tests.

Few would argue that aspirin should be banned. At the other extreme, most would agree that anabolic steroids should not be used. These are hormones which encourage the building-up of muscles. They do, however, have harmful side-effects, as well as being unfair, unless used by all competitors. Anabolic steroids were, at one time, taken widely by 'heavyweights', such as throwers and lifters. One noticeable effect is the appearance of masculine characteristics on 'female' athletes.

The first F.A. dope test took place after the Swindon v Brentford F.A. Cup tie on 24th November 1979, but the results were withheld until the end of the season. What would have happened if Swindon's players (who won 4-1) had been taking drugs, and had gone on to win the Cup?

Gambling

Predicting the results of sports events is bigger business even than the sports themselves. Horse racing and soccer attract the largest amount of gambling. So vital are the football pools felt to be, that a 'Pools Panel' meets to invent results if more than 25 league games are postponed.

A careful check of pools forecasts in newspapers suggests that they are no more accurate than any random method—sticking in pins, cathedral cities, birthdays etc.— of predicting drawn matches. If there are 8 drawn games, the odds against predicting them all are over 1000000000 to 1!

Sport is not responsible for gambling, nor for the money which people lose in betting on sports results. Social scientists are, however, interested in people's behaviour in this area. How many people, for example, gamble on football pools who would never 'back' a horse? How many horse-racing punters would never go in for other kinds of gambling?

Soccer Crowds

Professional sport, as well as affecting players' attitudes and attracting gambling, is entertainment for large numbers of people.

Whenever large numbers of people are packed into a small space, there are problems. For soccer crowds, the problem has always been the

safety of the actual ground. Crush barriers have occasionally collapsed, with injuries and deaths, the worst example being at Ibrox Park, Glasgow on 2nd January 1971, when 66 were killed and 145 injured.

Scenes like the estimated 200000 at the first Wembley F.A. Cup Final in 1923, and the 83000 watching Moscow Dynamo at Stamford Bridge in 1945, will never be repeated. The trend is towards seating rather than standing on terraces, with safety as a key factor.

In the 1970s, the main problem was the behaviour of some of those attending league soccer matches in Britain. Scenes like those shown were particularly associated with the visits of certain clubs.

Here is an eye-witness account of crowd disturbances at Carrow Road, Norwich on 2nd April 1977:

> 'Manchester United were coming to Norwich. After previous crowd trouble at their matches, all their games were all-ticket. Unable to get a seat I was offered a choice of standing at the River End (uncovered) or in the Barclay Stand (covered). Knowing that the 'Red Army' had been allocated half of the latter I chose to risk the weather.
>
> Despite being certain of entry I arrived at the ground a little after 2 p.m. and took up a place high in the River End. As well as a good view of the pitch, I had a clear view of the Barclay Stand at the opposite end of the ground. Before the match, there was much shouting, singing and chanting from the rival 'supporters' in the Barclay Stand. At one point, fighting broke out and innocent fans climbed on to the pitch to escape trouble as the police attempted to restore order.
>
> Manchester United manager Tommy Docherty went across in an unsuccessful attempt to calm the 'Manchester mob'. Finally, order was restored and the match began.
>
> Norwich won 2-1; during the match there was little trouble.
>
> The real problem started with the final whistle. Some of those aggrieved at losing seemed determined to take it out on the ground itself. Panels at the back of the stand were pulled away, broken and hurled at police, Norwich supporters, anything. Two youths even climbed onto the roof and began to demolish that—until one of them fell.
>
> The ground and road outside were afterwards like a battlefield. Many people were injured, including about 20 policemen, and there were several arrests.
>
> The City of Norwich returned to peace, wondering whether the trouble had been worsened by:
>
> (a) advance publicity that there would be trouble
> (b) a BBC film crew following Manchester United supporters
> (c) the result of the match.'

SPORT AND SOCIAL SCIENCE

Terrible pitch, but terrific floodlights!

Extra: Floodlights

To most people, sport is an opposite to work. It provides relaxation, a chance to concentrate on something other than life's problems. Much sport therefore takes place on outside normal working hours. For outdoor sports in winter this means that some kind of artificial lighting is necessary.

There are three scientific questions to answer:

(i) How bright is the lighting?
(ii) How much energy is required?
(iii) How will the lighting affect players' vision?

One problem with lighting is the 'inverse square law'. This means that doubling the distance from a light gives only one quarter of the brightness. For example:

Distance, m	1	2	3	4	5	6	10	12
Brightness, lumens	144	36	16	9	5.76	4	1.44	1

For a minimum-size soccer pitch, four sets of floodlights have to each illuminate $1250 Cm^2$. To prevent dazzle, damage and disturbance, the lights must be fairly high above the pitch. The inverse square law means that a large brightness is needed, consuming a lot of power. A professional ground may have to comply with special regulations, such as those of U.E.F.A. Southampton F.C. have ninety-six mercury halide lamps giving a total of over 150kW.

Players' vision will be affected as the ability to see detail is dependent on light intensity. Colour vision, too, is affected by both strength and colour of lighting. (In dim light, vision is restricted to shades of grey only.) The table shows the effects of coloured lighting. Try it for yourself.

Actual colour	Coloured Light					
	Red	Green	Blue	Yellow	Cyan	Magenta
Red	red	black	black	red	black	red
Green	black	green	black	green	green	black
Blue	black	black	blue	black	blue	blue
Yellow	red	green	black	yellow	green	red
Cyan	black	green	blue	green	cyan	blue
Magenta	red	black	blue	red	blue	magenta

Colours of objects under different coloured lighting

Soccer—the traditional 'working man's sport' in Britain—has been played under floodlights since 1878, although the first floodlit Football League match was not played until 1956, at Fratton Park, Portsmouth. Other sports followed, the most recent being cricket. The breakaway World Series Cricket introduced floodlit cricket to Australia. Restricted length evening matches became popular, and soon reached the 'official' game. Innovations such as a white ball, coloured pads and black sightscreens were found to improve visibility for the players.

References and Further Reading

Alexander, R. McN., *Animal Mechanics*, Sidgwick and Jackson, 1968
*Buchanan, D. *Greek Athletics*, Longman, 1976
Daish, C.B., *Ball Games*, English Universities Press (now Hodder and Stoughton), 1972.
Daish, C.B., *The Physics of Ball Games*, English Universities Press (now Hodder and Stoughton), 1972.
*Diagram Group, *Rules of the Game*, Bantam, 1976
Dyson, G., *The Mechanics of Athletics*, 7th Edn., Hodder and Stoughton, 1977
*Gillet, C., *All in the Game*, Penguin, 1971
Gray, J., *Animal Locomotion*, Wiedenfeld and Nicholson, 1968.
Lindsey, B.I., Sports Records as Biological Data, *J. Biol. Ed.* (1975) 9, 86-91.
McNab, T. (Ed.), *Modern Schools Athletics*, University of London Press (now Hodder and Stoughton), 1970.
*McWhirter, N., *Guinness Book of Records*, Guinness, 1978
Muckle, D. S. & Shepherdson, H., *Football Fitness and Injuries*, Pelham, 1975
*Nuffield Foundation, *Revised Nuffield Biology Text 2*, Longman, 1975
*Nuffield Foundation, *Nuffield Secondary Science, Theme 3*, Longman 1971.
*Nuffield Foundation, *Working with Science* units: *Sport, Football, Slimming*, Longman Resource Unit, 1977-78
*Page, R.L., *Man and Machines*, Pergamon, 1975
Page, R.L., *The Physics of Human Movement*, Wheaton, 1978
Physiotherapy, Sports Medicine Issue (1976) 62, 8, 245-265
Royal College of Physicians, *Smoking and Health Now*, Pitman Medical, 1972
Tozer, M.D.W., Investigations into Sporting Skills, *School Science Review*, (1972) 187, 54, 236-244.
*Tyler, M. (Ed.) *Encyclopaedia of Sports*, Marshall Cavendish, 1975

*indicates suitable for use by pupils

Appendix A: A Scheme for CSE Mode 3

Objectives

On completion of the course, students should be able to

1. describe and use the techniques of science appropriate to the study of sports;
2. manipulate and interpret data produced by such techniques and from sports results;
3. describe the physical factors involved in the flight of balls, and other sporting artefacts;
4. describe the anatomical and physiological basis of human activity, and its limitations;
5. assess the effects of practice on performance;
6. define the skills necessary for proficiency in various sports;
7. discuss the psychosocial aspects of sport as a factor in human ecology.

Syllabus

1. Techniques.
   (a) Measurement of length, mass, time, force, acceleration, speed, velocity, power, centre of gravity. SI units.
   (b) Analysis of movement by flicker-book, ticker-timer, multi-flash and stroboscopic photography, film, video-tape.
   (c) Significance of errors in measurement. Approximations.
   (d) Data handling; use of tables, graphs. Simple statistics.

2. Ball Games.
   (a) Physics of ball flight; momentum, friction, flow, drag, spin, rebound. Effects of ball texture on flight.
   (b) Types of stroke/shot and their effects.
   (c) Composition, manufacture & properties of playing surfaces.
   (d) Flight of objects other than balls.

3. Mechanical sports.
   Physical factors involved in motor-racing, sailing, gliding.

4. Athletic sports.
   (a) Anatomy and physiology of
      (i) skeleton: structure and properties of bone, cartilage, tendons, ligaments; limbs as levers; joints;
      (ii) muscle: types, properties, antagonism; effects of training;
      (iii) nerves: reaction time, reflexes, co-ordination, sight, hearing, balance;
      (iv) circulation;
      (v) respiration: anaerobic and aerobic; lactic acid; haemoglobin;
      (vi) excretion: salt and water balance, sweating;
      (vii) hormones: adrenaline and insulin;
      (viii) temperature regulation.

   (b) Physiology of exercise. Measurement of fitness.
   (c) Sports medicine. Injuries, first-aid, diet.

5 Skills.
   (a) Analysis of techniques.
   (b) Effects of practice.
   (c) 'Time and motion' studies.

6 Sport in Society.
   (a) Attitudes to sport. Gambling. Football pools, prediction.
   (b) Spectator sports, crowd behaviour.
   (c) Cheating.

Appendix B: Ideas for Project Work

Many of these ideas for projects were originated and/or carried out by the author's own pupils. Most of them can be modified to suit other sports.

1. Factors affecting the sprint start. Use of ticker-timer, electronic timing. Measurement of force, velocity, acceleration, impact time.

2. Centre of gravity studies. Analysis of centre of gravity movement during simple gymnastics exercise. Use of photographic techniques and/or manikin.

3. Long jump. Comparison of jumps produced by different approach speeds. Effect of horizontal 'bar' to encourage higher jump. Can carrying weights (Greek-style) really make a difference?

4. Swimming. Does training with a weighted belt improve performance under normal conditions?

5. Badminton. Analysis of shuttlecock flight by means of stroboscopic photography. Is there a difference between nylon and feathered shuttles? Measurement of impact force.

6. Baton speed during relay races. How fast does baton move compared with runners? What happens at take-overs?

7. Views and attitudes. Use of questionnaire to find opinions. Results analysed by age group, male/female etc.

8. Soccer data. What actually happens during a match? For how long is the ball 'in play'? What factors affect the size of crowds at League matches? Is there a relationship between players' numbers and goals scored? Would an earlier change from goal average to goal difference have made any real difference?

9. Can cricketers learn to bowl with the 'wrong' hand? Effects of results- and technique-dependent practice.

10. Left-handers. Analysis of Wimbledon entry to find if high proportion of left-handers. Proportion (and effects) of left-handed batsmen in test matches.

11. Show jumping. How do the skeleton and muscles of the horse compare with Man? Analysis of a jump. Balance of rider.

12. Damage. Manufacture and strength of damaged equipment. Study of types of damage and their causes.

Appendix C: Examination Questions

1. Give reasons for each of these practices:
   (a) Warming-up before playing or competing.
   (b) Putting on a track-suit (or similar clothing) when resting after exercise.
   (c) Not eating immediately before exercise.
   (d) Not going into bat directly from a dark pavilion.
   (e) Taking salt tablets after strenuous exercise.
   (f) Bending the knees when landing on the floor from a trampoline.

2. The amount of blood flowing through different organs can be measured during exercise, and compared to the value at rest. Typical results are:

Organ	Amount of blood flow compared to rest			
	Rest	Light exercise	Moderate exercise	Heavy exercise
BRAIN	1	1	1	1
HEART	1	1.5	2	2.5
MUSCLES	1	3	8	11
SKIN	1	2	3	1.5
KIDNEYS	1	0.8	0.7	0.5
OTHERS	1	1.1	1	0.8

   (a) Which organ has a constant blood supply?
   (b) Apart from the muscles, which organ shows an increase throughout?
   (c) Which organ shows a decrease throughout?
   (d) Why do the muscles need such a vast increase in blood supply?
   (e) During exercise, the amount of adrenaline in the blood increases greatly.
      (i) Where is adrenaline produced?
      (ii) What effects does it have on the body?
      (iii) Which substance has effects in some ways opposite to adrenaline?

3. (a) Draw a sketch of the bones in the arm. Label your drawing with names of the bones.
   (b) Which two muscles work the elbow joint?
   (c) Explain the differences between tendons and ligaments. Add one of each to your drawing.
   (d) Which injuries may the arm suffer?
   How would they be treated?

4. Several weightlifters had their 1976 Olympic medals taken from them.
   (a) Which drug had they taken?
   (b) What effects does it have on the body?
   (c) Why is it forbidden?

5

Time, s	Force, N	
	Front foot	Rear foot
0	0	0
0.1	750	1110
0.2	700	0
0.3	800	0
0.4	1000	0
0.5	0	0

The table shows the forces applied by each foot to the starting blocks during a sprint start.
  (a) Draw a graph to show the relationship between time and force for each foot singly, and both feet added together.
  (b) Describe, in words, what happens to the forces.

6 The table gives the dimensions of three different rackets. Draw a scaled sketch of each, and identify them.

Racket	A	B	C
Overall length, cm	65	69	69
Length of shaft, cm	41	47	38
Width of head, cm	20	19	23.5

7 In a training session, an athlete repeatedly lifts a 100N load from the floor to his shoulders.
  (a) If his shoulders are 1.5m from the floor, how much work does he do in a single lift?
  (b) How much work does he do in 100 lifts?
  (c) If the 100 lifts take 5 minutes, how much power does he develop?
  (d) Why should all such weight training be supervised?

8 The 1968 Olympic Games were held in Mexico City. The 5000m, 10000m and marathon were all won by Ethiopians.
  (a) Suggest a reason for this.
  (b) Why were they not at an advantage in the sprints and field events?
  (c) What would you expect to find if you could examine their blood closely?
  (d) How can this change in the blood be brought about?

9 (a) Draw a sketch of a section through the eye.
     Label: cornea, lens, iris, pupil, retina, blind spot.
  (b) What changes take place in the eyes as a sportsman watches a ball coming towards him?
  (c) Explain these phrases:
     (i)   having a good eye;
     (ii) keeping your eye on the ball;
     (iii) getting your eye in.

10 (a) Tennis courts may be made, among other materials, of grass, clay or concrete.
      Which, if any,
     (i)   gives a fast bounce; dries quickly?

(ii) gives a fast bounce; dries slowly?
(iii) gives a slow bounce; dries quickly?
(iv) gives a slow bounce; dries slowly?
(b) In the 1976 Olympic Games there were several accidents in the distance running events. What reason was given for this?

11 Two rugby players each took 10 kicks at goal every day for a week. Each kick was taken from the same place, but player R used his right foot, player L his left foot. Their successes are shown in the table:

Day		1	2	3	4	5	6	7
Goals	R	5	4	3	4	5	6	7
kicked	L	3	2	4	4	4	5	6

(a) Draw a graph to show these results.
(b) (i) On which day was L more successful than R?
(ii) On which days did R score with 50% of his attempts?
(c) Indicate whether each of these statements is true, false, or if there is not enough information to decide.
(i) All right-footed kickers are better than all left-footed kickers.
(ii) Performance improved with practice for both R and L.
(iii) R was more successful than L.
(iv) On day 8, R kicked 8 goals.
(v) Using two players is more accurate than using two whole teams.

12 Explain why
(a) A bowler (cricket) polishes one side of the ball.
(b) Squash players have a long 'knock-up' before play.
(c) Snooker players put chalk on their cues.
(d) Golfers don't use smooth balls.

13 Each of these statements about ENERGY is incorrect. Write a corrected version of each statement.
(a) Protein provides more energy per kg than any other food.
(b) The SI unit of food (and other kinds of) energy is the Watt.
(c) A pork chop contains as much energy as is used up in running up a flight of stairs.
(d) The conversion of chemical energy to kinetic energy takes place in the liver.
(e) When a golf ball flies through the air, it has both potential and light energy.
(f) Lifting 10 N a height of 5m needs 2 J of energy.

14 The drawings are traced from photographs of three stages in a golf shot.

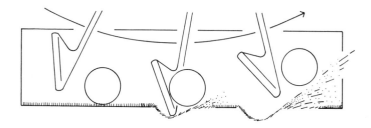

(a) What kind of club is being used?
(b) State three ways in which the original photographs may have been taken.
(c) Will the ball as drawn spin clockwise, anti-clockwise, or not at all?
(d) How will it bounce on landing?

15 Footballers are often spoken of as having 'cartilage operations'.
(a) Draw a healthy knee joint, showing cartilage.
(b) What is the function of cartilage?
(c) How may it be damaged?
(d) What is done in such a 'cartilage operation'?

16 (a) What is the average resting body temperature?
(b) Why is it wrong to refer to it as 'normal'?
(c) What happens to body temperature during exercise?
(d) How is a relatively constant temperature maintained?

Index

acceleration	50, 99
adrenaline	75, 87
aerobic respiration	81, 90, 96
aerofoil	100
anaerobic respiration	81, 90, 96
alveoli (air sacs)	84
Astroturf	107, 111
athletics	91 et seq.
jogging	122
timekeeping	33
see also individual events	
back-spin	60
balance	66
balls	51 et seq.
bounce	51-2
cricket	53
force on	45
surfaces	52-3
bar charts	29-31, 36
bicycle ergometer	85-6
blood	91, 96
bones	24
fractures	115-8
bounce	51-2
spin	60
surfaces	106
breathing	83-4
carbon dioxide	84
cartilage	24, 115
centre of gravity	15 et seq.
cycling	100
high jump	93
hurdling	91
cheating	124
classification of sports	8
colour vision	73, 127
control experiment	2, 113-4
cramp	92
cricket : ball	53
floodlit	127
left/right-hand	49
pitch	47, 106
swing/spin	61
darts	112
discus	94-5
dislocation	116
distance running	91
drag	53
motor racing	98-9
rowing	103
rugby	63
drugs	27, 125
ear : balance	66
hearing	74
efficiency	88-9
energy : conversion	46
food	86-7
kinds of	46
motor racing	98
muscles	79, 98
work	77
ergometer	85-6
exercise	79, 113-4
experiment	1, 113-4
eye	71-2
film	39
first aid	117-8
fitness	76
smoking	120
flicker-book	39
flight : gliding	100-1
manpowered	104-5
floodlights	127
food	24-5, 87
football : see soccer, rugby	
footwear	107
forces : balls	45, 50
car	98-9
glider	101
impact	56
sailing	102
fractures	115-6
friction : ball	54-5
footwear	107
ice	108-9
tyres	108
gambling	125
gliding	100-1
golf	56, 58-60, 64
graphs	29-31, 36
logarithmic	38
hammer throw	94-5
Harvard step test	82, 113
hearing	74
heart	82
height	11
high jump	17-18, 94
histograms	29-31, 36
hockey	49, 112
hormones	27, 75, 87-8, 92
hurdling	91
ice	108-9
impact force	56, 63
infections	119
information	2-4, 28 et seq.
injuries	115-8
insulin	87-8
javelin	94-5
joints	23, 43, 116
jogging	123
keys	9
lactic acid	81, 96
left-handed	49
levers	43-4
lift	98 et seq.
ligaments	24, 66, 116

INDEX

logarithmic graphs	38
long jump	92
lungs	83-4
Magnus effect	58
manikin man	18
man-powered flight	104-5
marathon	91
mass : ball	50, 56
body	11
weight	77
menstruation	27
momentum	49-50
motor racing	98-9
muscles : antagonistic	43
breathing	83
energy	79
injury	114
receptors	66
warming-up	119
nerves	68-9
Newton's laws	45
Olympic Games (Mexico)	95-6
oxygen	84, 89
blood	91, 96
debt	90
smoking	120
pacemaker (heart)	82
photography	40
pitches	106 et seq.
power	77
practice	112-3
predicting	4
press-ups	78-9
professional	123-4
proprioceptors	66
pull-ups	78
pulse rate	76-7, 113-4
pupils (eye)	73
questionnaires	5
reaction time	69-70
receptors	66
records	31-2, 37-8
red blood cells	96
reflection	61-2
reflex	67-8
respiration	81
results tables	29
right-handed	49
rowing	103
rugby	50, 62-3
running : analysis	44
distance	91
jogging	123
marathon	91
sprinting	44-5, 90
tracks	106-7, 110, 111
sailing	102
senses	65
shape	21-2
shot putt	94-5
SI units	28
skating	109
skeleton	24
skiing	109, 111
smoking	119
soccer : crowds	125-6
floodlights	127
money	124
Newton's laws	45
pitch	47, 109
pools	125
professionalism	123-4
spin	60-1
sport : classification	8
definition	6-7
sprains	116
sprinting	44-5, 90
squash	62
step test	82, 113
step-ups	77-8
stereoscopic vision	71
streamlining	53, 63
stroboscope	41
surface : area of body	20-1
ball	53
playing	105 et seq.
swerve	61-2
swimming	1-2, 45
swing	61
tables, results	29
'Tartan' surface	107
temperament	22-3
temperature : body	20, 81
squash ball	62
warming-up	119
tendons	24, 66, 117
tennis : force on ball	56
left/right hand	49
playing surfaces	105
practice	112
racket	57
spin on ball	60
throwing events	94-5
thrust	98-9
timekeeping	33
top-spin	60
tracks (running)	105-6
training	113-4
turbulence	53
tyres	107
units	28
velocity	50
video-tape	39
vision	71-3
warming-up	119
wear on pitches	110
weight	11
winter sports	108-9
work	77

Acknowledgements

Neither the original course, nor this book, would have been possible without the co-operation and encouragement of the staff and pupils of Archway School, Stroud. Particular thanks are due to the staff of the P.E. and Science Departments, especially Derek Barnes.

I should also like to thank David Cooper for his excellent illustrations, my wife for patient encouragement and for reading the typescript, and all friends and colleagues who have offered advice and comment.

The author and publishers would like to thank the following companies and institutions for permission to reproduce illustrations in this book:

All Sport Photographic Ltd. (Pages 11 Piggott, 13 Borzov, 14 Juantorena and Press, 27 Pele and Owens, 33 lower, 35, 81, 91, 95, 100, 103, 117, 126); David Finch – Photos (Page 14 Starbrook); Dendix Brushes Ltd. (Page 110); Dunlop Sports Company Ltd. (Page 51, 52 lower left); Goodyear Tyre and Rubber Co. (Great Britain) Limited (Page 108); London Express News (Page 105); Mitre Sports Ltd (Page 52 upper left and lower right); Running Magazine/West Highland News Agency (Page 123 Saville); Sport and General (Press Agency) Ltd (Page 13 McBride); Sporting Pictures (UK) Ltd (Pages 11 Alexeev and Ali, 13 Assinder, Connors and Grieg, 14 Cunningham, 27 Ashe and Cawley); Sports Council (Page 122); Timex Corporation (Page 33 upper).